Studies in Computational Intelligence

Volume 690

Series editor

Janusz Kacprzyk, Polish Academy of Sciences, Warsaw, Poland
e-mail: kacprzyk@ibspan.waw.pl

About this Series

The series "Studies in Computational Intelligence" (SCI) publishes new developments and advances in the various areas of computational intelligence—quickly and with a high quality. The intent is to cover the theory, applications, and design methods of computational intelligence, as embedded in the fields of engineering, computer science, physics and life sciences, as well as the methodologies behind them. The series contains monographs, lecture notes and edited volumes in computational intelligence spanning the areas of neural networks, connectionist systems, genetic algorithms, evolutionary computation, artificial intelligence, cellular automata, self-organizing systems, soft computing, fuzzy systems, and hybrid intelligent systems. Of particular value to both the contributors and the readership are the short publication timeframe and the worldwide distribution, which enable both wide and rapid dissemination of research output.

More information about this series at http://www.springer.com/series/7092

Peter Simon Sapaty

Managing Distributed Dynamic Systems with Spatial Grasp Technology

 Springer

Peter Simon Sapaty
Institute of Mathematical Machines
and Systems
National Academy of Sciences
Kiev
Ukraine

ISSN 1860-949X   ISSN 1860-9503   (electronic)
Studies in Computational Intelligence
ISBN 978-3-319-84405-3   ISBN 978-3-319-50461-2   (eBook)
DOI 10.1007/978-3-319-50461-2

Printed on acid-free paper

This Springer imprint is published by Springer Nature
The registered company is Springer International Publishing AG
The registered company address is: Gewerbestrasse 11, 6330 Cham, Switzerland

*To my grandson Vanya with true belief that
the world he so much loves and needs can be
secured and improved by this book*

Preface

This is the third book on high-level ideology, model, and technology for solving tasks in large distributed networked systems by active scenarios navigating and matching them in parallel. The other two being Mobile Processing in Distributed and Open Environments (1999) and Ruling Distributed Dynamic Worlds (2005), both from John Wiley & Sons.

The naturally arising question: What changed from the time of previous publications, and why this new book may be needed now? At least five reasons can be named.

Reason 1. Changed is the world as a whole, with its dynamics grown enormously for the last decade, lavishly fed by numerous conflicts and crises including international terrorism, ethnic, religious and military conflicts, endless floods of refugees, economy collapses too. To withstand this dynamics, much updated and even completely different views on the problems and their solutions are needed, in the area of information and control models and technologies too.

Reason 2. After the second book's publication, the developed ideology, methodology and technology were severely tested at many world events, both civil and military, to evaluate their potentials for solving hard practical tasks in distributed dynamic systems. And the proposed approach pretty much survived and was of real interest to different audiences. The feedback and experience gained helped to improve the technology and especially its core language in comparison with the previous book publications.

Reason 3. The participation in different world events with very different topics and very dissimilar audiences helped to witness uniformity of basic system principles in all areas. And this encouraged an attempt to link our work with some fundamental concepts like General System Theory, System Dynamics, Gestalt Psychology, also new trends like Memetics and Human Terrain; stimulating in this respect was our presentation at Gestalt Psychology Congress in Germany. It became clear that moving to higher system levels can often drastically improve solutions at lower levels too.

Reason 4. The previous technology versions were prototyped using special software communication channels between computers. But the current version can use any existing media communications, and can be easily installed in popular electronic devices which, communicating under the spreading spatial scenarios can help to solve different social problems within or without any borders.

Reason 5. Many conference and journal publications appeared during the last decade in relation to this technology and applications. It would be useful to collect at least most important of them and present in extended and improved form within a single volume like a book.

The mentioned above, altogether, have inspired the author to prepare this book in hope it can be useful and enjoyable by readers. Wishing them luck in digesting this material, with readiness to provide any additional consultation and information.

Kiev, Ukraine   Peter Simon Sapaty

Acknowledgements

The following persons and organizations contributed to the development of ideas and experience expressed in the current book.

Roland Vollmar, Institute of Theoretical Informatics, Karlsruhe, Germany, as the author's supervisor within Alexander von Humboldt Foundation (AvH) award and related projects carried out at the universities of Braunschweig, Karsruhe, and Oxford UK. His high scientific and moral standards and ardent support of novel concepts proved crucial in the development of ideas preceding this book.

Werner Zorn, Hasso-Plattner-Institut of the University of Potsdam, Germany, with whom the author supervised Intelligent Network Management project at the University of Karlsruhe financed by Siemens Nixdorf. High-level solutions for management of distributed computer networks were developed, related to this book too.

Lubomir Bic, School of Information and Computer Science, University of California, Irvine, USA, and Son Vuong, University of British Columbia Canada, who provided support of the author's works on parallel and distributed computing and hosted demonstration of the developed WAVE system at UCI and UBC.

Masanori Sugisaka, ALife Robotics, Japan, with whom the author has long and fruitful cooperation in the area of Artificial Life and Robotics, from working at the Oita University to organizing international ALIFE conferences to the grant of Japan Society for the Promotion of Science (JSPS), also having common publications mentioned in the book.

Stephen Lambacher, School of Social Informatics, Aoyama Gakuin University in Tokyo Japan, a great linguist with whom the author discussed many original ideas preceding this book, in direct contacts when worked together at the University of Aizu-Wakamatsu, and afterwards by email.

Robert Finkelstein, Robotic Technology Inc., USA, a great roboticist with whom the book related ideas were frequently discussed for more than a decade, some resulted in common publications mentioned in the book. Exchanges of opinion on memetics, human terrain, psychology, philosophy, social engineering, politics, system management, military to civilian conversion, etc. were inspiring.

Jürgen Kriz, University of Osnabrück, a great philosopher, who supported the author's ideas on holistic vision of large distributed systems and their relevance to Gestalt psychology, expressed in the book and presented at international Gestalt Theory Congress.

Bob Nugent, a retired US Navy Commander, with whom the author had meetings and useful discussions on advanced command and control, especially in maritime and underwater area, which inspired the author to show exemplary robotic scenarios in the book confirming advantages of the technology developed.

Very productive was cooperation within editorial boards of a number of international scientific journals which also organized invited publications of the book related papers. Special thanks to Svitlana Tymchyk and Natalia Karevina, Mathematical Machines and Systems journal (ISSN 1028-9763) of National Academy of Sciences in Ukraine, for friendship and lasting support. Similar thanks to Journal of Computer Science & Systems Biology, JCSB (ISSN: 0974-7230), International Journal of Advanced Research in Artificial Intelligence (IJARAI) of Science and Information (SAI) Organization (ISSN: 2165-4069, 2165-4050), and International Relations and Diplomacy (ISSN: 2328-2134).

The author is always grateful to his father Simon, a known biologist in the past, who researched in the area of photosynthesis of plants and lectured at Agricultural Academy in Ukraine, for taking his son from the early childhood and for all life into the scientific world of light, life, and growth. The current book has emerged from this wonderful world too.

Contents

Chapter 1
Introduction

1.1 What Is This Book About?

This book is *about distributed systems of different natures, and how to create, analyze, and manage them.* Such systems may consist of many physical, virtual or combined parts and interconnections and communications in between, which can be physical or information-based too.

Despite great successes in global communications where potentially any human or technical component of the system covering any area, the whole world including, can immediately communicate with any other one, the *world tissue remains largely distributed* and based on *direct local relations and communications* between individual components which altogether form a variety of organizational and operational structures.

The book is investigating and teaching of how to *create, modify, analyze, and support these structures* at a variety of system levels and their mixtures while grasping them mentally and technologically as a whole and organizing both efficient and effective behavior.

1.2 Offering Technological Response to World Challenges

We are witnessing a dramatic change in the character of national and international activity, especially in crisis and conflict areas, with engagement of asymmetric, unconventional, and hybrid methods and solutions to withstand the emerging problems and threats. The latter may simultaneously involve economy, ecology, international relations, ethnicity, culture, law, religion, sports, etc., defense and military too, covering both physical and virtual spectra. And the responding solutions may need to be multidimensional and highly integral in order to succeed.

© Springer International Publishing AG 2017
P.S. Sapaty, *Managing Distributed Dynamic Systems with Spatial Grasp Technology*, Studies in Computational Intelligence 690,
DOI 10.1007/978-3-319-50461-2_1

Within this global and often alarming context, the book describes a novel ideology and supporting Spatial Grasp Technology (SGT) for integral management of large distributed dynamic systems of both civil and defence orientation. The approach, developing further the invented model of parallel management of distributed systems expressed in the European patent [1] and previous books and book chapters [2–8], is based on a high-level Spatial Grasp Language (SGL) expressing solutions in physical, virtual, executive and combined environments in the form of active self-evolving and self-propagating patterns spatially matching the systems to be created, modified, controlled, and managed.

Within SGL, the communicating interpreters of SGL can be installed in key system points which may be in great numbers (millions to billions) and represent equipped humans, robots, laptops, smart phones, smart watches, smart sensors, etc. Operating under holistic and gestalt-inspired scenarios in SGL, initially injected from any points, these systems can be effectively converted into goal-driven spatial machines (rather than computers as dealing with physical matter and objects too) capable of responding to numerous challenges caused by the rapidly growing world dynamics in the 21st century.

1.3   Background and History

1.3.1   Creating Distributed Networks

The development history of SGT has more than four decades, having started with the creation of citywide heterogeneous computer networks, well before the internet [9]. Different computers in this project were working in very different high-level algorithmic languages but had to cooperate in order to solve complex combined analytical-numerical tasks [10–12] as neither computer could do this alone satisfactorily. Some computers directly emulated in hardware the language called ANALYTIC which worked with analytical expressions rather than numbers, generating formulae as a result, while the other, central computer, operated in a compiled FORTRAN within the numerical domain, and the two were incompatible with each other, even in principle.

1.3.2   Using Program Mobility

Instead of creating some new and very high level language covering the both languages with subsequent complex interpreted or compiled implementation for working on all computers of the network, a radically different solution had been found [13]. Within it, the formulae obtained in the ANALYTIC computer were propagating to the FORTRAN computer being implanted into a FOTRAN textual framework to become right-hand arithmetic expressions of the full scale FORTRAN programs to produce numerical results. The latter, when obtained in the FORTRAN computer, in their turn, were embedded into textual ANALYTIC framework and returned to the first computer as normal ANALYTIC programs to get new formulae again, and so on.

This solution, extremely simple and efficient, was actually the first known case where computers in a network were exchanging not just data but the whole programs in high-level languages. The methodology and practice of freely moving executive programs in computer networks carrying operations and control was called, years later, as *mobile agents*.

1.3.3   Active Graphs and Networks

Another line of our research at that time was aimed at finding parallel and fully distributed solutions on general graphs and networks, where each graph node could reside in a different computer with assigned local task related procedures doing something in the node and communicating with the neighboring nodes along the graph links [14–18]. After the thorough investigation, it appeared possible to find such local nodal activities which when communicating with other ones on the graph's topology could solve the graph and network problems as a whole and without central resources, and these local activities were very different for different tasks. The main problem was with preparation of the whole network for such parallel processing, especially for large and physically distributed graphs, and the supplying of its nodes with compiled local activities for each particular task. Another problem was global control of termination of such solutions in the distributed environment, which also restricted this at first sight promising approach for practical use.

1.3.4   Mobile Programs in Active Networks

Influenced by the abovementioned solutions of moving active program code rather than data carrying with itself operations and control and the activity of networks self-solving complex graph problems, a groundbreaking solution was found associating conceptually graph or network nodes not with local activities to be compiled and loaded for each for each task but rather with a *universal functionality* in the form of interpreter from a special high level language. The programs in such language could freely *propagate through the networks* and bring, at runtime, the needed operations and control into graph nodes for the solution of concrete tasks [19–21].

This language interpreter was able to solve *any graph and network problem* and need not loading each time for any new task, being rather a standard part of the distributed environment with same copies associated with different system components. The moving language, originally called WAVE [22–29], expressed and pursued a completely different distributed computing philosophy, psychology and culture, while expressing solutions of graph and network problems often orders of magnitude shorter and simpler than usual.

As an elementary example, let us imagine we want to build the shortest path tree of a distributed network starting from a node and leading to all other network nodes. The finding of such a tree in a fully distributed system by traditional approaches it not a trivial task requiring programming and putting certain procedures into each node, establishing communication and synchronization with other nodes, also setting global control over the network to determine when the distributed solution is found, and son.

In WAVE, you could just start from what to be the root node and move in parallel through the network links carrying growing distance from the start and redefining the direct predecessor node if the distance brought to a node is smaller than the registered before. And *this is the whole algorithm* occupying less than a string of WAVE code, in comparison to about a hundred of strings in other languages expressing traditional algorithms.

If you wanted to find the shortest path from the tree starting node to any other node, you could start from the latter and cycle though the already built tree (recorded in the network structure by predecessor addresses in nodes) back to its root—another half a string in WAVE! The main secret is that 99% of what was explicitly written in existing algorithmic languages was *covered implicitly and automatically* by universal internal interpretation mechanisms of the wavelike language.

This waves-based spatial approach represents a certain background for the current book too, with its Spatial Grasp Language (as a remote heir of WAVE and its subsequent derivatives) having many additional and powerful both forward and echo network navigating features revealed in the book.

1.3.5   International Support

The following organizations provided financial and spiritual support for the development, testing, and international reporting of the ideas preceding the development of SGT, especially in the areas of artificial intelligence, intelligent management of distributed systems, and robotics:

- National Academy of Sciences of Ukraine;
- USSR Academy of Sciences and National Council of Artificial Intelligence;
- Slovak Academy of Sciences and its International Basic AI Laboratory in Bratislava;
- Alexander von Humboldt (AvH) Foundation, Siemens Nixdorf company and universities of Braunschweig, Karlsruhe, and Siegen in Germany;

- Universities of Oxford and Surrey, Defense Research Agency and Ericsson company in the UK;
- University of British Columbia in Canada;
- Japan Society for the Promotion of Science (JSPS) and universities of Oita and Aizu-Wakamatsu.

1.4   The Book Organization

The material of the rest of this book is organized as follows.

Chapter 2 contains some philosophical and theoretical background of the presented approach which includes biology-inspired *General Systems Theory*, *System Dynamics* with computer simulated multi-loop network models, *Brain Waves and Consciousness* phenomena, *Gestalt Psychology* grasping the whole first with parts having sense only in the context of the whole, *Memetics* similar to genetics but in information and social areas, also currently dominant *Interoperability* principles of organization of civil and especially military systems and the necessity and urgency of higher organizational levels like *Over-operability* coined by the author.

Chapter 3 describes the *Spatial Grasp Model* based on distributed spatial patterns self-matching the distributed worlds and its key element—recursive *Spatial Grasp Language*, SGL, in which all such patterns are expressed in a parallel wavelike mode, with unlimited scenario mobility in distributed spaces. SGL directly operates with physical, virtual, and executive worlds and their any combinations, also with pure computations within the same universal recursive syntax which allows the language to be extended easily for new classes of distributed applications. Elementary scenarios in SGL and main ideas of its networked interpretation are presented too.

Chapter 4 provides full *specification of SGL* (with scenarios in it called "grasps") and semantics of main constructs including constants, different types of stationary and mobile spatial variables, and universal operational and descriptive units called "rules". Due to recursion, constants and rules can themselves be arbitrary grasps, allowing for flexible spatial scenarios at different system levels and their mixtures inducing parallel forward and echo processes, generalization of distributed situations for making local and global decisions, and direct processing of remote data with results retained in the navigated environments or returned to the starting points as higher-level knowledge.

Chapter 5 describes main *spatial mechanisms* of SGL with the help of progress points or "props" as integration of emerging control with the reached locations in navigated spaces and use of improvised 3D graphics. Different mechanisms activated by the SGL rules for both breadth and depth evolution of spatial scenarios are explained and exhibited, including parallel, sequential, synchronous, asynchronous, alternative, sliding, repetitive, cyclic, looping, slinging, whirling, and other variants. All these, especially if used together, are allowing for the overall vision, investigation, analysis, and control of large distributed spaces, with great power and clear and compact linguistic expressions.

Chapter 6 reveals general structure and organization of *SGL interpreter* which can be networked with other similar interpreters for solving tasks in parallel and distributed way. Main data structures of the interpreter, its functional processors, and their interactions are described which themselves may work together as an integral and parallel system, depending on implementation. The scenario history-based *forest-like track system* of the distributed interpreter allows it to serve as spatial support and control backbone providing overall awareness and processing of the returned data, while delivering new scenario code and data to the world positions reached previously, also maintaining and updating distributed information resources with their removal if not used anymore.

Chapter 7 shows how an exemplary *distributed network-shaped world* can be created from scratch, modified, and managed in SGL. Different shape-forming distributed techniques are presented, with the created world navigated by mobile agents interacting with each other and with the world. The latter can be active too, with its nodes informing each other to any depth and breadth on the active agents moving through them. This world is subsequently globally controlled by runtime introduction of higher, supervision, layers which can change the overall functionality and behavior. Also shown is how a virtual world can be converted into a physical one and autonomously move in parts or as a whole in distributed space, preserving overall integrity and shape. The whole model is written within the same SGL syntax which makes it transparent, compact, and simple.

Chapter 8 is devoted to solving some known *theoretical graph and network problems* where each graph or network node may be in a separate location/computer, with links connecting nodes in the same or in different computing facilities. Fully distributed and parallel solutions in SGL are shown for different paths between nodes, shortest path trees and shortest paths that can be used for different purposes including optimal delivery of physical substance between two points, weak and strong components (like articulation points and cliques) with measures to modify networks influenced by them, different strategies of forming templates for finding arbitrary graph structures in arbitrary networks, also some specific structures that may be of particular interest. The presented scenarios can effectively involve into operation and cooperation all network resources which may be in great numbers and spread worldwide.

Chapter 9 is for solving different *societal problems* in distributed systems, naming typical ones and introducing a new trend in the field called Human Terrain which stems from military but may have important applications in civil areas too. Many problems can be investigated using social networks effectively represented and processed in SGL. The shown cases are influenced by theoretical tasks from the previous chapter, they also consider finding topological center of a community and regular checking physical distance between such centers of different communities, alarming if a certain threshold reached. Tracing movement of elderly people in large urban environments with emergency medical support, also distributed relief solutions after crises or disasters, including guided collective evacuation from a disaster zone, are exhibited too.

Chapter 10 investigates application of the developed approach for the support of *automated Command and Control* (C2). It starts with purely semantic scenario in a distributed environment with fully automatic C2 and overall system organization provided by formal networked interpretation of SGL. A detailed C2, if needed, can also be expressed and implemented in SGL which is shown for a fleet in maritime environment. Other described cases relate to distributed tracing of cruise missiles having complex routes which can be effectively followed and controlled by mobile spatial intelligence spreading through sensor networks; ideas of a possible European missile defense with flexible hierarchical C2 capable of handling many targets simultaneously; also comparing traditional components-based battlefield management with its semantic equivalent in SGL, the latter much simpler and more compact thus improving capability of runtime modification in unpredictable situations.

Chapter 11 describes different *collective robotic scenarios* with multiple unmanned units operating under spatial C2 generated by formal network interpretation of SGL. These combine randomized swarm movement with regularly updated group's topological centre and stemming from it hierarchical C2, the latter collecting and impacting multiple targets throughout the area covered; the use of cooperating robots in hospital service; multiple robotic mapping of unknown spaces; battling forest fires by robotic swarms; costal waters cooperative patrol; collective finding of oil spill centre in water covered surfaces; collective robotic attack against a fleet in maritime space, swarm against swarm aerial operation; and the use of multiple robotic units in forestry and agriculture with various strategies of space coverage. All such solutions in SGL are very compact and can be composed and updated at runtime, with robotic collectives operating autonomously and without external control.

Chapter 12 concludes the book which cultivates spatial vision and finding of effective solutions on highest levels of organization of distributed dynamic systems, ranging from social systems to critical infrastructures to dynamic battlefields to massive collective robotics, allowing, if needed, to describe, manage, and simulate any other system levels too. And this can avoid the use of different languages for different system levels with space and time consuming interfaces in between. Along with general conclusions, the chapter highlights contribution and relation of the developed spatial grasp approach to the system theories and trends briefed in Chap. 2. It also summarizes, with corresponding references, application areas of particular importance where SGT was presented at different international events. Prospects of installation of the latest technology version on existing platforms and future plans are discussed too.

References

1. P.S. Sapaty, A distributed processing system, European Patent No. 0389655, European Patent Office (1993)
2. P.S. Sapaty, Logic flow in active data, in *VLSI for Artificial Intelligence and Neural Networks* (1991), pp. 79–91
3. P. Sapaty, *Mobile Processing in Distributed and Open Environments* (Wiley, New York, 1999)
4. P. Sapaty, *Ruling Distributed Dynamic Worlds* (Wiley, New York, 2005)
5. P.S. Sapaty, Distributed technology for global control, in *Informatics in control, automation and robotics*, Vol. 37 of the series Lecture Notes in Electrical Engineering (2009), pp. 3–24
6. P.S. Sapaty, Meeting the World Challenges with Advanced System Organizations, in *Control, Automation and Robotics*, vol. 85 of the series Lecture Notes in Electrical Engineering (2011), pp. 29–46
7. P.S. Sapaty, Grasping Spatial Integrity in Distributed Unmanned Systems, in *Informatics in Control Automation and Robotics*, vol. 85 of the series Lecture Notes in Electrical Engineering (2011), pp. 79–97
8. P.S. Sapaty, The World as distributed brain with spatial grasp paradigm, in *Intelligent Systems for Science and Information*, vol 542 of the series Studies in Computational Intelligence (2014), pp. 65–85
9. A.T. Bondarenko, S.B. Mikhalevich, A.I. Nikitin, P.S. Sapaty, Software of BESM-6 computer for communication with peripheral computers via telephone channels, in *Computer Software*, vol. 5 (Institute of Cybernetics Press, Kiev, 1970) (in Russian)
10. P.S. Sapaty, A Method of organization of an intercomputer dialogue in the radial computer systems, in *The Design of Software and Hardware for Automatic Control Systems*, Inst. of Cybernetics Press, Kiev (1973) (in Russian)
11. J.G. Grigorjev, V.P. Karpus, L.I. Pristupa, P.S. Sapaty, Management of a dialogue in the MIR-2–BESM-6 system, in *Proceeding of Republic Conference on Hardware and Software for Management of Dialogue in Computer Systems*, Kiev (1973) (in Russian)
12. A.T. Bondarenko, S.B. Mikhalevich, P.S. Sapaty, Intercomputer dialogue in high-level languages, in *Proceeding of Republic Conference on Hardware and Software for Management of Dialogue in Computer Systems*, Kiev (1973) (in Russian)
13. P.S. Sapaty, On possibilities of the organization of a direct intercomputer dialogue in ANALYTIC and FORTRAN languages, Publication No. 74–29, (Institute of Cybernetics Press, Kiev, 1974) (in Russian)
14. P.S. Sapaty, Solving branching and cycling tasks on multiprocessor systems, in *Proceeding USSR Academy of Sciences*. Technical Cybernetics, No. 1 (1974) (in Russian)
15. P.S. Sapaty, Organization of computational processes in distributed heterogeneous computer networks, Ph.D. Dissertation, Institute of Cybernetics, Kiev (1976) (in Russian)
16. P.S. Sapaty, On efficient structural implementation of operations on semantic networks, in *Proceeding USSR Academy of Sciences*. Technical Cybernetics, No. 5 (1983) (in Russian)
17. P.S. Sapaty, Active information field as a model for structural solving of tasks on graphs and networks, in *Proceeding of USSR Academy of Sciences*. Technical Cybernetics, No. 5, (1984) (in Russian)
18. P.S. Sapaty, Solving tasks on semantic networks and graphs by active distributed structures. in *Proceeding of 3rd International Conference Artificial Intelligence and Information-Control Systems of Robots*, Smolenice (Elsevier Science Publishers B.V., North-Holland, 1984)
19. P.S. Sapaty, A wave approach to the languages for semantic networks processing, in *Proceeding of International Workshop on Knowledge Representation. Section 1: Artificial Intelligence*, Kiev (1984) (in Russian)
20. P.S. Sapaty, A wave language for parallel processing of semantic networks. Comput. Artif. Intell. **5**(4) (1986)

21. P.S. Sapaty, The wave approach to distributed processing of graphs and networks, in *Proceeding of International Working Conference Knowledge and Vision Processing Systems*, Smolenice (Nov 1986)
22. P.S. Sapaty, WAVE-1: a new ideology of parallel processing on graphs and networks, in *Proceeding of International Conference Frontiers in Computing*, Amsterdam (1987)
23. P.S. Sapaty, The WAVE-1: A new ideology and language of distributed processing on graphs and networks. Comput. Artif. Intell. **5** (1987)
24. P.S. Sapaty, in *Parallel Processing for Knowledge Representation*, ed. by C. Jesshope. Infotech State of the Art Report on Parallel Processing (Pergamon Press, England, 1987)
25. P.S. Sapaty, WAVE-1: a new ideology of parallel processing on graphs and networks. Future Gener. Comput. Syst. **4** (1988)
26. P.S. Sapaty, The WAVE model for advanced knowledge processing. Report No. OUEL 1803/89 (University of Oxford, England, 1989)
27. P.S. Sapaty, The WAVE machine project, in *Proceeding of IFIP Workshop on Silicon Architectures for Neural Nets*, St. Paul de Vence, France, 28–30 Nov 1990
28. P.S. Sapaty, The WAVE paradigm, in *Proceedings of JICSLP '92 Post-Conference Joint Workshop on Distributed and Parallel Implementations of Logic Programming Systems*, Washington, D.C., 13–14 Nov 1992
29. P.S. Sapaty, A brief introduction to the WAVE language. Report No. 3/93, Faculty of Informatics (University of Karlsruhe, 1993)

Chapter 2
Some Theoretical Background

2.1 Introduction

The Chapter contains some theoretical background of the approach presented in the book including *General Systems Theory* inspired by biology and introducing open systems concept, *System Dynamics* with computer-simulated multi-loop network flow models, *Gestalt Psychology* with the phenomenon of grasping the whole first while treating parts in the context of the whole rather than vice versa, *Memetics* as an analogue to genetics but in social and information areas, and *Brain Waves and Consciousness* investigated on biological and psychological levels. Also analyzed are currently dominant *Interoperability* principles in organization of large civil and military systems, revealing their insufficiency and confirming necessity of introduction of higher-order organizational levels. The latter, symbolically called *Over-operability*, should orient on holistic, integral and global-goal-driven system organizations and applications which could withstand challenges emerging in large distributed dynamic systems.

2.2 General Systems Theory

This theory after Ludwig von Bertalanffy [1–6] is based on similar general conceptions and viewpoints which have emerged and evolved in various disciplines. In the past, science tried to explain observable phenomena by reducing them to communication and interplay of elementary units which are independent from each other. New approaches are based on what is termed "wholeness" where systems of various orders cannot be understood, even in principle, by investigation of their parts in isolation. Conceptions of this nature have appeared in all branches of science, regardless of which types of objects are being studied (i.e. inanimate things, living organisms, social phenomena, etc.).

© Springer International Publishing AG 2017
P.S. Sapaty, *Managing Distributed Dynamic Systems with Spatial Grasp Technology*, Studies in Computational Intelligence 690,
DOI 10.1007/978-3-319-50461-2_2

Fig. 2.1 Open systems
concept

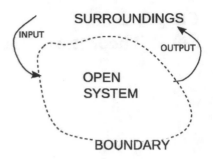

General Systems Theory (GST) is an interdisciplinary field of science and the study of the nature of complex systems, a framework within which any group of objects that work together in concert can be analyzed and described. The theory attempted to provide alternatives to conventional models of organization, with applications to numerous areas of study, emphasizing *holism* over reductionism, *organism* over mechanism. There are many instances where identical principles were discovered in different fields, and these principles can be analyzed and classified regardless of the origin and nature of the systems.

One of central themes of Bertalanffy's work is discovery and analysis of *nonlinear relationships* between system variables, where a small change in one variable may cause a significant change in another one. Another important contribution of GST is the theory of *open systems* (see Fig. 2.1) which, existing within certain boundaries, have input and output flows representing exchanges of matter, energy or information with their surroundings.

The open systems concept appeared to be particularly relevant to *living organisms*. Bertalanffy maintained that the conventional formulation of physics is, in principle, inapplicable to the living organisms, which keep themselves in a continuous inflow and outflow with the surroundings, also in building up and breaking down of components. They do not exist, so long as alive, in a state of chemical and thermodynamic equilibrium but maintain so-called *steady state* which is distinct from the latter.

In *social sciences*, Bertalanffy believed that general systems concepts were applicable too, criticizing classical atomistic conceptions of social systems. The GST encouraged for new developments from sociology, to anthropology, economics, political science, and psychology among other areas.

The holistic tune dominates the whole theory stating that the "progress is only possible by passing from a state of undifferentiated wholeness to differentiation of parts". Bertalanffy considered the entire world (symbolically in Fig. 2.2) as an *organization*, with biosphere as a whole, and this vision can drastically change traditional categories, our kind of thinking, also influence practical activity in different areas of life.

Fig. 2.2 The whole world as
an organization

2.3   System Dynamics

Fundamental works have been carried out by Jay Forrester on analysis of complex
systems (urban, industrial, world as a whole) and their detailed computer simulation
[7–10]. It has become clear that complex systems are *counterintuitive*, where cor-
rective actions are often ineffective or even adverse in the results. In complex
systems the cause and the effect are often not closely related in either time or space,
and their structure is not a simple feedback loop.

In reality, complex system has a *multiplicity of interacting feedback loops*, and
its internal rates of flow are controlled by *nonlinear relationships*. Complex systems
are usually of high order, with many system states or levels. They may contain
positive feedback loops reflecting growth, and negative, goal seeking ones. In the
complex system the cause may lie far back and away from symptoms, and the
causes are usually found not in prior events but rather in the *structure and policies*
of the system.

Figure 2.3 shows the simplest possible feedback system using just two symbols:
a stock, and a flow.

The stock may be considered as an accumulation, integration or level, depending
on the terminology from different fields, and the flow changes the amount in the
stock. The flow is determined by a statement that tells how the flow is controlled by
the value of the stock in comparison to a goal. All systems and everywhere consist
of these two kinds of concepts, stocks and flows.

Another, a bit more complex, model introducing a second order loop is depicted
in Fig. 2.4.

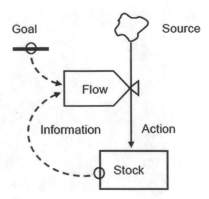

Fig. 2.3 Simplest feedback system

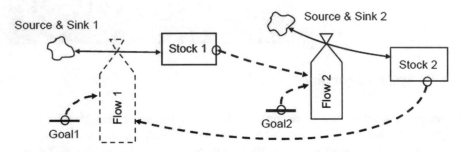

Fig. 2.4 Feedback system with second order loop

More complex network flow models (with hundreds of nodes and thousands of links in between) incorporated interrelations between such features as population, capital investment, natural resources, pollution, and agriculture at different system levels, etc. These levels took into account changes in the rates of flow in and out, such as birth and death rates which were causing the increase and decrease of population.

The detailed definition of the model states how each rate of flow is assumed to depend on the levels of population, natural resources, capital investment, capital devoted to agriculture, pollution, and so on. These models incorporated many nonlinear relationships.

The basic components of the network flow models with brief explanation of their meanings are shown in Fig. 2.5.

Using the system dynamics model, a computer simulation can show how a system, as described for each of its parts and links in between, would behave. Given a set of starting conditions, the simulation can calculate and plot the results that unfold through time.

The great uncertainty with mental models arises from their inability to anticipate the consequences of interactions between parts of a system, which may be in great

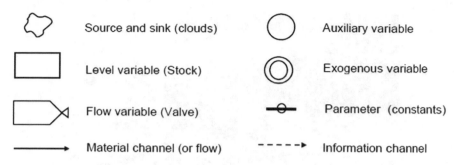

Fig. 2.5 Basic elements of network flow models

numbers and complex relationships. This uncertainty may be *totally eliminated* in computer simulation models. Forrester showed that the proper simulation is becoming a powerful procedure for clarifying many issues in large dynamic systems which are often counterintuitive for the human mind.

2.4   Gestalt Psychology

Gestalt is both philosophy and psychology term which means *unified whole* [11–14]. It refers to visual perception theories developed by German psychologists at the beginning of the past century. These theories tried to understand and describe how people can organize visual elements into groups or unified wholes under certain conditions and principles, maintaining our meaningful perceptions in an apparently chaotic world.

For the gestaltists, "Gestalten" are not the sums of aggregated contents erected subjectively upon primarily given pieces. Instead, we are dealing with wholes and whole–processes possessed of inner intrinsic laws. Elements are determined as parts by the intrinsic conditions of their wholes and are to be understood as parts relative to such wholes.

Gestalt psychology was founded by German thinkers Max Wertheimer, Wolfgang Kohler and Kurt Koffka and focused on how people interpret the world. According to Gestalt psychology, *the whole is different from the sum of its parts*. Based upon this belief, Gestalt psychologists developed a set of principles to explain perceptual organization, or how smaller objects are grouped to form larger ones. These principles are often referred to as the "laws of perceptual organization."

The school of gestalt practiced a series of theoretical and methodological principles that attempted to redefine the existing approach to psychological research. This was in contrast to investigations developed at the beginning of the 20th century, based on traditional scientific methodology, which divided the object of study into a set of elements that could be analyzed separately with the objective of reducing the complexity of this object.

Fig. 2.6 Perceiving partial images as a whole

The word "Gestalt" is usually translated as form, although it might be better understood as *organized structure*, as opposed to heap, aggregate, or simple summation. A traditional perception example is shown in Fig. 2.6 depicting a Dalmatian dog sniffing the ground in the shade of overhanging trees.

The dog is not recognized by first identifying parts (like feet, ears, nose, tail, etc.) and then inferring the dog from those component parts. Instead, the dog is *perceived as a whole*, all at once.

By Gestalt psychology, the mind understands external stimuli as a whole rather than the sum of their parts. The wholes are structured and organized using grouping laws, which are called *laws* or *principles*. The most important laws with their names, meanings and simple explaining graphics are briefed below. (These laws deal with the sensory modality of vision but there are also analogous laws for other modalities including auditory, tactile, gustatory and olfactory.)

- **Law of Proximity** (Fig. 2.7) states that when individuals perceive an assortment of objects, they often first perceive objects that are close to each other as forming a group. According to the law of proximity, things that are near each other seem to be grouped together.
- **Law of Similarity** (Fig. 2.8) states that elements within an assortment of objects are perceptually grouped together if they are similar to each other. This similarity can occur in the form of shape, color, shading or other qualities.
- **Law of Closure** (Fig. 2.9) states that individuals perceive objects such as shapes, letters, pictures, etc., as being whole when they are not complete. Specifically, when parts of a whole picture are missing, our perception fills in the visual gap.
- **Law of Symmetry** (Fig. 2.10) states that the mind perceives objects as being symmetrical and forming around a center point. It is perceptually pleasing to

Fig. 2.7 Proximity example

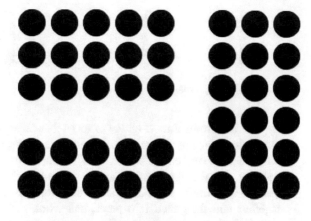

Fig. 2.8 Similarity example

Fig. 2.9 Closure example

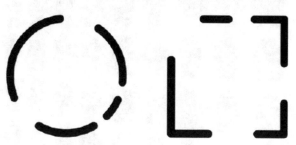

divide objects into an even number of symmetrical parts, and when two symmetrical elements are unconnected the mind perceptually connects them to form a coherent shape.

Fig. 2.10 Symmetry example

- **Law of Common Fate** (Fig. 2.11) states that objects are perceived as lines that move along the smoothest path. Experiments using the visual sensory modality found that movement of elements of an object produce paths that individuals perceive that the objects are on.
- **Law of Continuity** (Fig. 2.12) states that elements of objects tend to be grouped together and integrated into perceptual wholes if they are aligned within an object. In cases where there is an intersection between objects, individuals tend to perceive the two objects as two single uninterrupted entities.
- **Law of Good Gestalt** (Fig. 2.13) explains that elements of objects tend to be perceptually grouped together if they form a pattern that is regular, simple, and orderly. This law implies that as individuals perceive the world, they eliminate complexity and unfamiliarity so they can observe a reality in its most simplistic form.

Fig. 2.11 Common fate example

Fig. 2.12 Continuity example

Fig. 2.13 Good gestalt
example

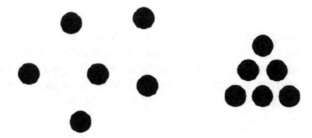

- **Law of Past Experience** (Fig. 2.14) implies that under some circumstances
  visual stimuli are categorized according to past experience. If two objects tend to
  be observed within close proximity, or small temporal intervals, the objects are
  more likely to be perceived together.
- **Figure/Ground** (Fig. 2.15) refers to the relationship between positive elements
  and negative space. The idea is that the eye will separate whole figures from
  their background in order to understand what's being seen. It's one of the first
  things people will do when looking at any composition.

Although Gestalt psychology and theory was a very general approach with great
potentials, most of the work on gestalt has been carried out in the area of perception
(the laws and examples mentioned above just confirming this).

In our true belief, the theory and practice of gestalt can be effectively used in
much broader sense and scale too, and especially for holistic global vision,

Fig. 2.14 Past experience
example

Fig. 2.15 Figure/ground
example

comprehension and proper management of large distributed dynamic systems of
most different natures, which is so important nowadays and which is the main
theme of this book. Establishment of broader and more universal gestalt laws
oriented on complex situations in large dynamic systems with proper technological
support would be particularly important too.

2.5 Memetics Versus Genetics

Memetics [15–17] is the theory of mental content based on an analogy with
Darwinian evolution. Being an emerging subfield of psychology, memetics is
considered as an approach to evolutionary models of cultural information transfer.
The *meme*, analogous to a *gene*, was conceived as a "unit of culture" (like idea,
belief, pattern of behaviour, etc.) which is "hosted" in the minds of one or more
individuals, and which can reproduce itself, jumping from mind to mind (as
symbolically shown in Fig. 2.16).

Memes consist of information which persists, propagates, and influences human
behaviour; they have an independent existence, self-replicate, and are subject to
selective evolution through environmental forces. They spread through the social
body similar to how genes spread through the biological body, forming the invisible
DNA of human society. And like a virus moves from body to body, memes move
from mind to mind.

Potentially, memetics can be effectively used to identify and target specific root
causes of challenging social problems in different areas. It may also be useful for
understanding the origin and evolution of modern humans, and to provide insight
into science, industry, and technology.

Critics of memetics, however, contend that some of its assertions are still
untested, unsupported or incorrect, even considering it as pseudoscientific dogma
and threat to the study of consciousness and cultural evolution. Some also say that
culture cannot be best understood by examining its smallest parts like memes, but
rather is pattern-like, comparable to an ocean current.

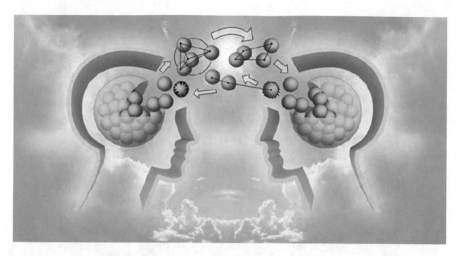

Fig. 2.16 Symbolic memes exchange

2.6   Brain Waves and Consciousness

Many things and activities in the world and universe are existing and proceeding, also considered as such, in the form of waves from, say, electromagnetic, radiation, gravitation, heat, sound, seismic and ocean waves to the waves of economic development [18], spiritual waves [19], democracy's waves [20], waves in linguistics [21], social waves [22], crime waves [23], immigrant waves [24] and many others like, for example, brain waves [25–27]—the latter to be discussed below in some details.

The human brain is a complex entity which is constantly at work, sending electrical signals, communicating, building new neural connections and so on. This electrical activity generated by the brain, also known as *brainwaves*, reflects our state of mind (as symbolically in Fig. 2.17). If we deepen our understanding of these brainwave frequencies, we can control our reality.

There are different kinds of brainwaves occurring at a specific frequency or pattern. Each brainwave has its own set of characteristics representing a unique state of consciousness. The different brain waves and the resultant states are usually classified as follows.

- **Beta** (12–30 Hz) are the brainwaves of our *normal waking consciousness*. It is associated with a heightened state of alertness, logical thinking, problem-solving ability, concentration, when the mind is actively engaged in mental activities. Most people spend their waking lives in a beta state.
- **Alpha** (8–12 Hz) brainwaves are slower in frequency as compared to Beta, which translates to a *highly relaxed state of awareness*. It is a normal brainwave pattern in people who are naturally relaxed and creative. Children tend to have much higher levels of alpha brainwaves than adults.

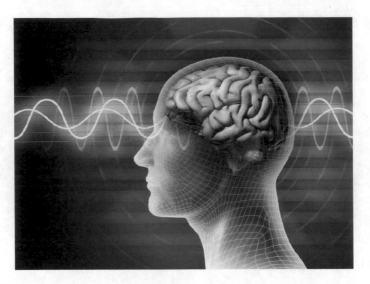

Fig. 2.17 Brain waves

- **Theta** (4–7 Hz) brainwaves occur during *deep relaxation and meditation*, light sleep or lucid dreaming. It is the realm of your sub-consciousness, where the mind is capable of profound insight, advanced intuition, and healing. Most children and teenagers have dominant theta brainwave patterns.
- **Delta** (0.5–4 Hz) waves are the slowest in frequency but are the highest in amplitude. Observed in *deep, dreamless sleep*, this frequency is the gateway to the universal mind and the collective unconscious, where information received is otherwise unavailable at the conscious level. It is a dominant brainwave of infants and even adults in deep sleep.
- **Gamma** (25–100 Hz) brainwave is the fastest frequency at which the brain functions, where an individual can experience *bursts of insight or high-level information processing*. It is the state of feeling that you can do anything.

People can alter their brainwave pattern in order to reach a desired level of consciousness. For example, if you can't get sleep at night when you feel stressed, you can synchronize your brainwave into the frequency that corresponds to sleep using sound. During meditation you reach a calm and relaxed state, corresponding to the alpha state of consciousness.

The concept of waves is also used in a study of higher levels of mental activity like, for example, *integral psychology* [28, 29], which is determined to embrace and unite all aspects of human consciousness under one concept, integrating ideas and models of consciousness, psychology, and therapy. The resultant psychological model includes *waves of development, streams of development, states of consciousness*, and the *self*. Under existing general consensus, neither mind nor brain can be reduced without each other, which means that both mind and brain need to be included in a *non-reductionistic way* in any integral theory of consciousness.

The idea of brainwaves on both biological and psychological levels can be exploited in a much broader sense and scale, covering such conceptions as self-awareness, self-analysis, self-organization, self-control, self-recovery, and many others. If supported by proper information technologies, it can also be effective for integral, holistic management of complex systems, not only localized but arbitrarily distributed in virtual and physical spaces too, converting them into a sort of holistic brain having a sort of spatial consciousness and pursuing global goals. These ideas and thoughts will be influencing the remaining material of the current book.

2.7 Interoperability Organizations and Their Weakness

Another system concept, widely used at present for organization and management of large distributed systems, especially military and international, is called *interoperability* [30–33], as a quality of a system with understood and clear interfaces to work with other systems without restrictions.

There may be different kinds of interoperability. *Syntactic interoperability* means that two or more systems are just capable to communicate with each other. *Semantic interoperability* supposes that beyond the ability to exchange information, different systems are capable of interpreting the exchanged information. *Cross-domain interoperability* occurs when different kinds of entities (which may be multiple social, organizational, political, legal, etc.) can work together for a common purpose.

Interoperability is currently the key principle for joint operations in both civil (Fig. 2.18) and military (Fig. 2.19) areas.

For example, NATO has been based on interoperability since 1949 when it was founded. Interoperability allows organisations of different nationalities and armed services to conduct joint peacekeeping operations. Interoperability is also dominant in any international relief missions acting after natural or manmade disasters.

Over the last several decades, the military has greatly benefited from the increased knowledge and capabilities provided by using computerized command and control systems. As this use has expanded exponentially, so has the need to integrate these systems. The breadth of computing technology at the component, functional, and mission level has complicated the issue of interoperability. By their nature, these disparate systems have varying levels of fidelity, granularity, quality and availability. The cost of establishing collaboration between these systems is typically high, and is complicated by differing organizational readiness levels, willingness, and technical ability to affect collaboration.

The need for translation of information and data to forms that are readable and interpretable by every unit has continuously challenged users of computer systems. Over time, the technologies employed to accomplish interoperability have evolved. Initially, and still prevalent today, one-to-one interfaces explicitly define how two systems interact. This type of approach works but does not scale. Other approaches,

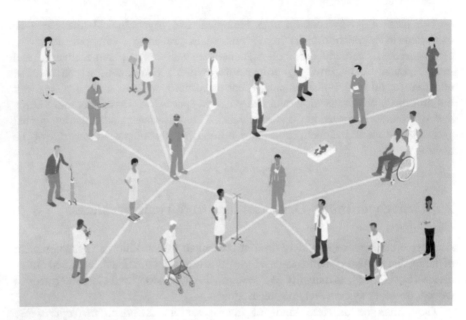

Fig. 2.18 Interoperability in a civil area

Fig. 2.19 Military interoperability

such as shared databases, common data repositories, and defined common standard messaging and interface formats, present solutions to some interoperability issues but are not panaceas. Each approach is appropriate in given circumstances only. Attempts to provide a single solution for all scenarios typically fall short due to technical challenges, adoption resistance, and funding availability.

Taking also into account the increased complexity of joint operations due to the growing world dynamics and emerging instability makes existing interoperability principles not sufficient to provide the needed overall awareness, integrity, and pursuit of global goals with runtime adjustment to new ones. The situations are often complicated by the necessity to operate in cyber-contested multi-dimensional spaces with high connectivity and inter-dependence, also with numerous actors having own, often quite different purposes and interconnections.

We are often witnessing failures of the currently dominant interoperability-based approaches and systems at national and international levels, especially where seemingly honest and noble intensions and actions of different players to improve complex situations in certain places on the globe lead to quite unexpected and even dangerous results. This mainly occurs by incapability of grasping the current and future whole of the problem properly, which is aggravated by the lack of appropriate scientific research, advanced system models and supporting technologies for its expression and management.

2.8   Over-Operability Versus Interoperability in System Organization

The current dominant interoperability organizations continue exercising *atomistic*, parts-to-whole philosophy of the system comprehension and implementation, criticized and rejected by holistic and gestalt approaches even a century ago. Interoperability is also very close to the atomistic vision of how human mind functions, like a "society of mind" [34], which proved its fundamental weakness and even incorrectness.

Let us briefly outline a possible alternative way to the design, formalization, implementation, and management of large distributed systems, which may help us overcome the weaknesses of interoperability organizations. Calling this *over-operability* [35, 36], also being gestalt-based [37, 38], the following ideas are the result of thorough investigation of how integral distributed systems should be planned, designed, and managed, also decades of practical experience in the field.

The usual sequence of design and implementation steps of large distributed systems may look like what is shown in Fig. 2.20.

Originally, the system idea and its expected functionality and capability emerge in a very general, holistic, indivisible and non-verbal form in a single human mind or close collective of such minds.

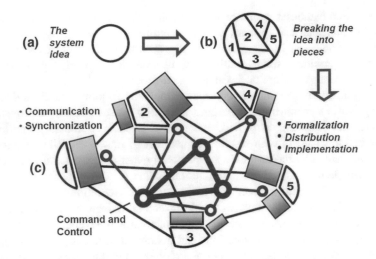

Fig. 2.20 Traditional steps in system design and implementation

Then this initial and "shapeless" idea, symbolically shown Fig. 2.20a just as a circle, is tried to be partitioned into individual chunks (shapeless too, but representing certain potential functionalities, still inseparable from the whole and from each other and having no individual meaning), each trying to be understood, detailed, redefined, and mentally studied further under the context of the whole, Fig. 2.20b.

This logical partitioning inevitably causes "swelling" of the problem's complexity, because each chunk needs autonomous definition, discussion, and investigation of its potential relations with the other chunks, which were not present and even imagined at the initial stage of Fig. 2.20a.

The next step is materialization of the defined parts and their relations with each other, using existing formal models, programming and simulation languages and technologies, as well as methods and capabilities of distribution in physical and virtual spaces. To make these parts working together as the whole system and believably within the original idea of Fig. 2.20a, we may need a good deal of their communication (with both physical matter and information) and synchronization, also a sophisticated command and control infrastructure generally distributed rather than concentrated (which may need introduction of additional physical and virtual entities or "chunks" too), as depicted in Fig. 2.20c.

The overhead with such distribution, communication, synchronization and control may be essential, even in some sense outweighing the original (considered "useful" and "pure") project functionality and complexity of Fig. 2.20a. But this overhead may need to be tolerated in any way, as parts of the distributed system may have to be located in particular physical and virtual points in order to serve local dynamic data there and other systems directly associated with these locations.

As can be seen, the real system formalization, description, and implementation (using existing philosophies, models, languages and technologies) can start only from the already developed distributed, interlinked, and synchronized stage of Fig. 2.20c (directly relating to what we called above as *interoperability*). And this can cause (and really does) numerous problems, some of which are as follows.

- It is often difficult to put the resultant distributed system with many interlinked parts *into compliance* with the initial idea representing the system as a whole, as in Fig. 2.20a, partially Fig. 2.20b too.
- The hopefully correctly assembled whole for the separate parts may have *quite different features* than expected, including unwanted and even dangerous ones.
- The resultant solution of Fig. 2.20c is *predominantly static*. If the initial idea of Fig. 2.20a changes, the whole system may have to be redesigned and reassembled, partially or even completely.
- Adjusting the already existing multi-component system to a new idea and new goals may be with high price and considerable loss of performance.

In this book, we will be introducing a special holistic or *over-operability* layer G shown in Fig. 2.21, which can help us to substantially reduce or even eliminate the problems mentioned above.

The establishment of such a layer, supported by special formal model and high-level universal language as opposite to informal one of Fig. 2.20b, may allow us to keep top system definition, description, and modification in a semantic, integral, compact, and very flexible form. The latter also allowing numerous traditional details of system partitioning, implementation, load distribution, and restructuring, especially at runtime, to be shifted partially or even completely to

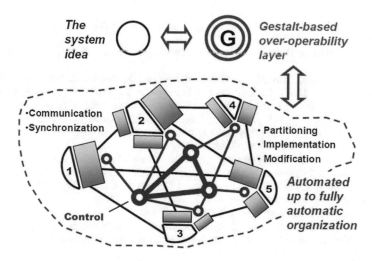

Fig. 2.21 Establishing over-operability layer for system creation and management

effective automatic level (which can massively use manned, unmanned and mixed components).

This over-operability layer will be the main theme of this book. Allowing us to effectively grasp large distributed systems on topmost level, it will be creatively using the described above ideas of *general system theory, system dynamics, gestalt psychology, memetics, brain waves,* and *interoperability,* among others. It will also be further investigating and developing the integral vision of large distributed dynamic systems originally expressed in the European patent [39] and previous books and book chapters [40–46].

2.9   Conclusion

Considered were some fundamental theories and novel approaches shedding the light on how large distributed systems of different natures, from technical to biological to social, are organized, behave and how they should be managed to fulfill complex objectives. The real shift should be done, first of all, from still dominating atomistic parts-to-whole philosophies and approaches, which were in doubt even a century ago, towards integral and holistic vision and management of large dynamic systems in order to provide increased power, overall awareness, and pursuit of global goals. And this over-operability needs to be supported by advanced system models and information and control technologies to move forward in reality and practice, which is the main goal and content of this book.

References

1. L. von Bertalanffy, *General System Theory Foundations, Development, Applications* (George Braziller, New York, 1968)
2. L. von Bertalanffy, The theory of open systems in physics and biology. Science **111**(2872) (1950)
3. L. von Bertalanffy, Theoretical models in biology and psychology. J. Pers. **20**(1) (1951)
4. D. Pouvreau, M. Drack, On the history of Ludwig von Bertalanffy's "general systemology", and on its relationship to cybernetics. Part I: elements on the origins and genesis of Ludwig von Bertalanffy's "general systemology". Int. J. General Syst. **36**(3), 281–337 (2007)
5. M. Drack, Ludwig von Bertalanffy's organismic view on the theory of evolution. J. Exp. Zool. (Mol. Dev. Evol.) **324B**, 77–90 (2015)
6. D. Pouvreau, On the history of Ludwig von Bertalanffy's "general systemology", and on its relationship to cybernetics - Part II: contexts and developments of the systemological hermeneutics instigated by von Bertalanffy. J. General Syst. **43**(2), 172–245 (2014)
7. J. Forrester, *Urban Dynamics* (Pegasus Communications, 1969)
8. J. Frrester, *World Dynamics* (Wright-Allen Press, 1971)
9. J. Frrester, *Industrial Dynamics* (Pegasus Communications, Waltham, MA, 1961), p. 1961
10. J. Forrester, Learning through system dynamics as preparation for the 21st century. 2009. Retrieved from 11 Jan 2009

11. J. Wagemans, J.H. Elder, M. Kubovy, S.E. Palmer, M.A. Peterson, M. Singh, R. von der Heydt, A century of gestalt psychology in visual perception: I. Perceptual grouping and figure-ground organization. Psychol. Bull. **138**(6), 1172 (2012)
12. M. Wertheimer, Untersuchungen zur Lehre von der Gestalt, I: Prinzipielle Bemerkungen. Psychologische Forschung (1922)
13. W. Köhler, Die physischen Gestalten in Ruhe und im stationären Zustand. Eine natur-philosophische Untersuchung. Braunschweig. Germany: Friedr. Vieweg und Sohn (1920)
14. K. Koffka, *Principles of Gestalt Psychology* (Lund Humphries, London, U.K., 1935)
15. R. Dawkins, *The Selfish Gene* (Oxford University Press, 1976)
16. B.J. Hancock, Memetic Warfare: The Future of War, Military Intelligence, PB 34–10-2, **36**(2) (April–June 2010)
17. R. Finkelstein, Tutorial: Military Memetics, in *Social Media for Defense Summit*, Alexandria, Virginia, Robotic Technology Inc., 24–26 Oct 2011
18. W. Kwasnicki, in *Waves of Socio-Economic Development—An Evolutionary Perspective*, ed. by T.C. Devezas, Kondratieff Waves, Warfare and World Security (IOS press, Amsterdam, 2006), ISBN: 1-58603-588-6
19. D. Hall, Making spiritual waves and breaking old patterns, http://www.mediate.com/ACRFamily/docs/HALL-MAKING%20SPIRITUAL%20WAVES%20AND%20BREAKING%20OLD%20PATTERNS.pdf
20. S.P. Huntington, Democracy's third wave. J. Democracy **2**(2), 12–34 (1991)
21. A. Francois, in *Trees, Waves and Linkages: Models of Language Diversification*, ed. by B. Claire, E. Bethwyn. The Routledge Handbook of Historical Linguistics (Routledge, 2014), pp. 161–189
22. J. Adams, E. Clemens, A.S. Orloff, Social Theory, Modernity, and the Three Waves of Historical Sociology, as the Introduction to J. Adams, E. Clemens and A. S. Orloff, eds. *Remaking Modernity: Politics, History and Sociology* (Duke University Press, 2004)
23. R. Di Tella, S. Galiani, E. Schargrodsky, Crime distribution and victim behavior during a crime wave, 13 Nov 2006. http://fundacaofhc.org.br/files/apresentacoes/1884.pdf
24. B. Bell, S. Machin, Crime and immigration: evidence from large immigrant waves in the UK, Feb 2010, http://cep.lse.ac.uk/seminarpapers/02–02-10-BB.pdf
25. J.L. Fannin, Understanding your brainwaves. White Paper, http://drjoedispenza.com/files/understanding-brainwaves_white_paper.pdf
26. J.L. Fannin, The seven keys to great meditation: a neuroscience perspective, http://drjoedispenza.com/files/the_seven_keys_to_great_meditation.pdf
27. L. Gabriel, Brain wave basics—what you need to know about states of consciousness, Thought Medicine. Exploring the Power of Mind from Science to Spirituality (2011), http://thoughtmedicine.com/2011/06/brain-wave-basics-what-you-need-to-know-about-states-of-consciousness/
28. K. Wilber, Waves, streams, states, and self: an outline of an integral psychology. Humanistic Psychol. **31**(2–3) (2003)
29. K. Wilber, *Integral Psychology: Consciousness, Psychology, Therapy* (Shambala Publications, Boston, Spirit, 2000)
30. Connecting health and care for the nation a shared nationwide interoperability, roadmap DRAFT Version 1.0. The Office of the National Coordinator for Health Information Technology, (2015), https://www.healthit.gov/sites/default/files/nationwide-interoperability-roadmap-draft-version-1.0.pdf
31. D. Allen, S. Karanasios, A. Norman, Information sharing and interoperability: the case of major incident management. Eur. J. Inf. Syst. **23**, 418 (2014). doi:10.1057/ejis.2013.8
32. M.P. Gallaher; A.C. O'Connor, J.L. Dettbarn Jr., L.T. Gilday, Cost analysis of inadequate interoperability in the U.S. capital facilities industry. Report, National Institute of Standards and Technology, p. iv. Retrieved 2012-04-19 (Aug 2004)
33. Interoperability for joint operations, NATO backgrounder (2006), http://www.nato.int/nato_static/assets/pdf/pdf_publications/20120116_interoperability-en.pdf

34. M. Minsky, *The Society of Mind* (Simon and Schuster, New York, 1988)
35. P. Sapaty, Providing Over-operability of Advanced ISR Systems by a High-Level Networking Technology, in *SMI's Airborne ISR*, Holiday Inn Kensington Forum, London, United Kingdom, 26–27 Oct 2015
36. P.S. Sapaty, Over-operability in distributed simulation and control. MSIAC's M&S J. Online, Winter Issue, **4**(2) (2002)
37. P. Sapaty, Gestalt-based ideology and technology for spatial control of distributed dynamic systems, in *International Gestalt Theory Congress, 16th Scientific Convention of the GTA*, University of Osnabrück, Germany, 26–29 March 2009
38. P. Sapaty, Gestalt-based integrity of distributed networked systems, in *SPIE Europe Security + Defence*, bcc Berliner Congress Centre, Berlin Germany (2009)
39. P.S. Sapaty, A Distributed Processing System. European Patent No. 0389655, European Patent Office (1993)
40. P.S. Sapaty, Logic Flow in Active Data, in *VLSI for Artificial Intelligence and Neural Networks* (1991), pp. 79–91
41. P. Sapaty, *Mobile Processing in Distributed and Open Environments* (Wiley, New York, 1999)
42. P. Sapaty, *Ruling Distributed Dynamic Worlds* (Wiley, New York, 2005)
43. P.S. Sapaty, Distributed Technology for Global Control, in *Informatics in Control, Automation and Robotics*, vol. 37 of the series Lecture Notes in Electrical Engineering (2009), pp. 3–24
44. P.S. Sapaty, Meeting the World Challenges with Advanced System Organizations, in *Control, Automation and Robotics*, vol. 85 of the series Lecture Notes in Electrical Engineering (2011), pp. 29–46
45. P.S. Sapaty, Grasping Spatial Integrity in Distributed Unmanned Systems, in *Informatics in Control Automation and Robotics*, vol. 85 of the series Lecture Notes in Electrical Engineering (2011), pp. 79–97
46. P.S. Sapaty, The World as Distributed Brain with Spatial Grasp Paradigm, in *Intelligent Systems for Science and Information*, vol. 542 of the series Studies in Computational Intelligence (2014), pp. 65–85

Chapter 3
Spatial Grasp Model

3.1 Introduction

The chapter describes Spatial Grasp Model (SGM) based on networked patterns self-propagating and self-matching the distributed networked worlds. It also explains main principles of its key element—recursive Spatial Grasp Language, SGL, in which all such patterns are expressed.

The matching patterns provide parallel wavelike coverage of distributed spaces, with waves not being a result of the system's activity as usual (like, for example, electromagnetic, heat, or brain waves), but rather behaving as *active substances themselves* bringing operations and control into the systems covered. This may remotely resemble the old technique called *mobile agents* [1, 2], which in our case have evolved from their traditional *atomistic meaning* and usage into the *holistic grasping of the world* and influencing its behavior globally.

SGL effectively operates with physical, virtual, and executive worlds, also their combinations within the same recursive syntax and semantics. This allows us to feel and effectively express *direct personal presence* in the worlds with mental movement to and between their physical and virtual points, which is very different from traditional programming psychology and styles. SGL, moreover, supports this mental presence and activity in a highly *parallel* manner, which may well *exceed* natural brain's capabilities in a variety of important applications.

Key details of SGM and SGL, examples of elementary programming in the language, and general ideas and concepts of SGL networked interpretation in distributed dynamic systems are presented in the chapter too. The development history of SGM and SGL and their use in different projects related to large distributed systems can be traced via the existing publications [3–49].

© Springer International Publishing AG 2017

P.S. Sapaty, *Managing Distributed Dynamic Systems with Spatial Grasp Technology*, Studies in Computational Intelligence 690, DOI 10.1007/978-3-319-50461-2_3

3.2 Spatial Grasp Model Key Issues

3.2.1 General Idea

In most general terms SGM can be explained as shown in Fig. 3.1.

Any needed impact on a distributed world is represented as an active spatial pattern rather than traditional program, sequential or parallel. This pattern, expressing top semantics and key decisions of the problem to be solved, applied in certain world points, spatially *grows, covers and matches* the world at runtime. The pattern can create, control and change parts of the world or its whole, with final results retained in the distributed space (as passive information or active self-organized systems, generally distributed too) or returned to the starting points as detailed data or high level knowledge, whatever required, for a further engagement.

3.2.2 Parallel Wavelike World Coverage

The organization and evolution of this matching pattern is based on *parallel wavelike coverage, grasping, or conquest* of distributed physical and virtual worlds, as symbolically shown in Fig. 3.2 (for a single source application) which, if needed, can be assisted by a hierarchical feedback control and return of results which may be remote.

The matching pattern evolves as a succession of arbitrary complex spatial operations, or *grasps*, each developing, potentially in parallel, from final virtual or physical points of the previous grasp. Invocation of new grasps may take place immediately upon reaching any final results by the previous grasps, or may need preliminary synchronization of their full completion and analysis.

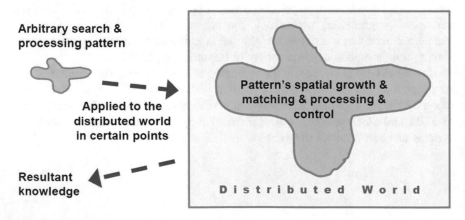

Fig. 3.1 Spatial grasp model main idea

Fig. 3.2 Controlled wavelike
space coverage

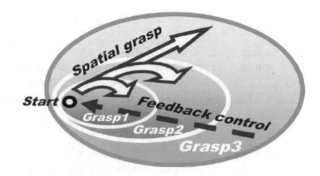

3.2.3   Navigation Pattern's Modification, Reduction, and Replication

The active body of the pattern can freely propagate in space during pattern's
evolution and world matching with omission of utilized parts if not needed any
more, which makes the patterns self-organized and independent of the systems they
are determined to match. For the symbolic case of Fig. 3.2, we may have the
following stepwise reduction of the pattern during its space propagation (semicolon
here used to separate successive grasps).

$$\text{Grasp 1; Grasp 2; Grasp 3} \rightarrow \text{Grasp 2; Grasp 3} \rightarrow \text{Grasp 3}$$

If a grasp results in multiple destinations, the whole remaining sequence of
grasps can be replicated and started in them in parallel. This replication may repeat
until the entire pattern is exhausted, as follows:

$$\text{Grasp 1; Grasp 2; Grasp 3} \rightarrow /after\,Grasp\,1/$$
$$(\text{Grasp 2; Grasp 3}),\ldots,(\text{Grasp 2; Grasp 3}) \rightarrow /after\,Grasp\,2/$$
$$(\text{Grasp 3},\ldots,\text{Grasp 3}),\ldots,(\text{Grasp 3},\ldots,\text{Grasp 3})$$

3.2.4   Spatial Grasp with Echo Processing

Described above was the *forward evolution* of matching patterns covering the
distributed systems and producing results inside of them. Another key feature of the
model is its *feedback echo processes* supported by powerful internal mechanisms
triggered by constructs called *rules* embracing the scenarios, which can spatially
process multiple results obtained by grasps in distributed spaces and return them to
the starting points if needed. For example, by writing

rule (Grasp)

we can first navigate, cover, and process the space by Grasp in a forward way, and then, within echo process, spatially collect, generalize, compare, process and return to the starting point any results obtained throughout the whole region covered, whatever large and remote, for a further use. In another example:

rule (Grasp 1; Grasp 2; Grasp 3)

the rule covers the whole combination of grasps, which may be any. A context-like rule may also guide and instruct the embraced grasps on how to propagate through the world and process it, like the following one adding creative features to the propagation:

create (Grasp)

A control rule can, for example, clarify how different grasps should operate together during the world coverage, say, either to succeed each other during advancement in space or be launched from the same point as sequential or parallel branches, as follows:

advance (Grasp 1, Grasp 2, Grasp 3)
sequential (Grasp 1, Grasp 2, Grasp 3)
parallel (Grasp 1, Grasp 2, Grasp 3)

A matching pattern may have any combination and hierarchy of grasps covered with rules, like the one shown below and in Fig. 3.3 (accompanied with a symbolic physical analogue). This provides virtually unlimited power of grasping, analyzing and processing of arbitrary spaces, both physical and virtual, with leaving results in the navigated system's body or bringing them back (this may just be a confirmation of total success or failure, or any other estimates) if needed.

rule 4(rule 1 (Grasp 1), rule 2 (Grasp 2), rule 3(Grasp 3))

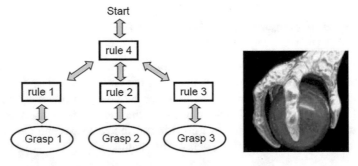

Fig. 3.3 Echo-based world grasping with symbolic physical analogy

We have explained only the most general ideas and features of the space navigation by self-matching, self-replicating, and self-modifying patterns leaving results in the navigated spaces or analyzing and returning them in echo processes. Any concrete development of such spatial operations may depend on a hierarchical control embedded into the patterns expressed by rules. This control can be arbitrarily complex where subsequent grasps may be deferred at dynamically appearing local or global control centers and broadcast to the points reached by other grasps only after thorough analysis and only if needed, thus optimizing communications and traffic in networks.

3.2.5   Multisource Matching

The matching pattern may also start from multiple sources and by different players, with same or different bodies evolving from them, as shown in Fig. 3.4, where constituent patterns, possibly originating at different times, can cooperate or compete in the distributed space or just ignore (i.e. being transparent to) each other.

The multisource organization of Fig. 3.4 in case of the same constituent patterns can be formally represented, if convenient, as a single source one similar to Fig. 3.2. This can be easily done with an additional grasp (e.g. Grasp 0) starting

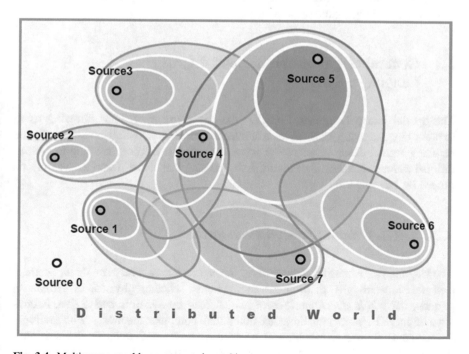

Fig. 3.4 Multisource world coverage and matching

from some additional source (like Source 0, which can also be one of Source 1 to Source 7) and moving initially to the other sources (with proper delays if needed) before applying of the remaining common pattern in them, as follows:

$$\text{Grasp 0; Grasp 1; Grasp 2; Grasp 3}$$

3.2.6   Combining Biological, Sociological, and Psychological Ideas

The briefed above pattern-based and feedback-controlled growing, spreading, coverage and matching ideology of dealing with large distributed systems effectively combines parallel spatial processes resembling the genes-, memes-, also waves-based development and growth of biological and social systems with holistic grasping of spatial images in them which may well relate to gestalt psychology (these system features and trends were discussed in the previous chapter).

This allows us, as will be shown on numerous examples throughout this book, to solve complex problems in a variety of distributed systems and on different stages: from parallel controlled propagation in space—to creation of self-organized distributed infrastructures—to providing overall awareness and sophisticated control by the latter—to modification of the created infrastructures by their navigation with other self-matching patterns—and so on.

3.3   General Organization of Spatial Grasp Language, SGL

The Spatial Grasp Language (SGL) is the key element of SGM allowing us to express any matching patterns in a universal, concise, and holistic way with the resulting formal descriptions called *scenarios* rather than programs due to nontraditional nature of SGL, also taking into account that these descriptions may be played by humans too.

3.3.1   SGL Orientation and Peculiarities

The language allows us to *directly move through, observe, and provide any actions and decisions* in fully distributed environments, whether physical or virtual. In general, *the whole distributed world*, which may be dynamic and active, is considered in SGL as a substitute to traditional computer memory, with multiple "processors" (humans, robots, any manned or unmanned units or devices, etc.) directly operating in it in a cooperative or competitive manner.

3.3.2 SGL Recursive Structure

SGL has a recursive structure with its top level shown in Fig. 3.5 which reflects the grasping nature of SGM matching patterns shown in Figs. 3.2 and 3.3 (with words in italics representing syntactic categories, square brackets showing an optional construct, braces indicating repetitive parts, and parentheses and comma being the language symbols).

This organization allows us to express any spatial algorithms, create and manage any distributed structures and systems, whether static or dynamic, passive or active, also solve any problem *in*, *on*, and *over* them, and this often can be expressed in a compact and unified way.

An SGL scenario, called *grasp*, applied from a certain world point can just be a *constant* representing the result explicitly or a *variable* containing data assigned to it previously, say, by other SGL scenario (or another branch of the current one) which has already visited this point. It can also be a *rule* expressing some action or definition, optionally supplied with certain parameters (enclosed in parentheses and separated by comma if more than one—generally represented as *grasp* again).

3.3.3 Constants

SGL constants can represent *information*, physical *matter* (physical objects including), self-identifying *custom* items (relating to information, matter or both), or *special* standard parameters or modifiers used throughout the language in different constructs. The final option generalizes constant as *grasp* again, potentially

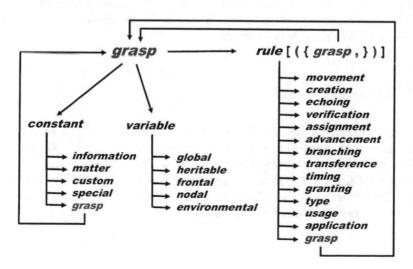

Fig. 3.5 SGL recursive syntax

allowing it to represent any objects within SGL syntax (passive or with embedded activities) and with any structures, for their further processing by SGL rules.

3.3.4 Variables

SGL variables, called "spatial", containing information and/or matter and supporting different features of distributed scenarios, can be stationary or mobile. They are classified as *global* (with overall common access), *heritable* (event-born and shared by subsequent events), *frontal* (accompanying scenario evolution and propagating in space, i.e. mobile), *nodal* (associated with visited world nodes and locally shared by different processes appearing in them), and *environmental* accessing external (navigated) and internal (language implementation) environments.

3.3.5 Rules

The rules, starting their influence in the current world positions, can be of different natures and levels—from local matter or information processing to full depth management and control. They can produce results residing in the same or other world positions. The results obtained and world positions reached by rules may become operands and/or starting positions for other rules, with new results and new positions (single or multiple) obtained after their completion, and so on. The rules are covering such language features as *movement, creation, echoing, verification, assignment, advancement, branching, transference, timing, granting, type, usage*, and *application*. The final rule's option, *grasp*, brings another level of recursion into SGL where rules may themselves be defined by the results of any SGL scenarios rather than mentioned explicitly.

3.4 More SGL Details

The SGL scenario can dynamically *spread* & *process* & *match* the world or its parts needed, with the scenario code capable of virtually or physically splitting, replicating, modifying, and moving in the distributed spaces being accompanied with transitional data. This movement can take place in single or multiple scenario parts dynamically interlinked under the overall control which is spreading and covering the navigated world too. More SGL details follow.

3.4.1 SGL Worlds

The language directly operates with:

- *Physical World* (PW), continuous and infinite, where each point can be identified and accessed by physical coordinates expressed in a proper coordinate system (terrestrial or celestial) and with precision given.
- *Virtual World* (VW), which is discrete and consists of nodes and semantic links between them, with both nodes and links capable of containing arbitrary information, of any nature and volume. VW can be considered as finite, but taking into account the rapidly growing world information and the internet growth may potentially be classified as infinite too. It may be hierarchically structured, with higher level nodes containing lower level ones together with links interconnecting them, and links themselves may also contain lower level nodes with their interconnections, and so on.
- *Executive world* (EW) consisting of active "doers" with communication possibilities between them. These may represent any devices or machinery capable of operating on the previous two worlds—including properly equipped humans, robots, mainframes, laptops, smartphones, intelligent sensors, etc. EW can also be hierarchical too, with higher level doers (say, groups, organization, or even societies) consisting of lower level ones down to separate individuals, with proper communications between doer nodes at different levels.

Different kinds of combination of these worlds can also be possible within the same formalism. For example, *Virtual-Physical World* (VPW) may not only be a mere mixture of the two worlds but also their deeper integration where individually named VW nodes can associate with certain PW coordinates and therefore exist in physical reality too. On the other hand, the whole regions of PW (of arbitrary shape and size) may have identifying virtual names, and this naming can be hierarchical. Another possibility is *Virtual-Execution World* (VEW), where doer nodes may have special names assigned to them and semantic relations in between, similarly to pure VW nodes. *Execution-Physical World* (EPW) can have doer nodes pinned to certain PW coordinates as, say, being stationary in these locations, and *Virtual-Execution-Physical World* (VEPW) can combine all features of the previous cases.

The main components of SGL (i.e. constants, variables and rules) have sufficient features and capabilities for expressing, processing and control of all these worlds.

3.4.2 How SGL Scenarios Evolve

Some more details on how SGL scenarios self-evolve in distributed environments, with the key issues following.

- SGL scenario is considered developing in *steps*, which can be *parallel*, with new steps produced on the basis of previous steps.

- Any step, including the starting one, is always associated with a certain *point* or position of the world (i.e. physical, virtual, executive, or combined) in which the scenario (or its particular part, as there may be many parts working simultaneously) is currently developing.
- Each step provides a resultant *value* (which may be single or multiple, also structured) representing information, matter or both, and a resulting control *state* (as one of possible states ranging by their strength, as shown later). This resultant state may be issued in the step's starting point or in other, reached, point (or points, which may be many).
- Different scenario parts may evolve from the same points in *ordered*, *unordered*, or *parallel* manner, as interdependent or independent steps.
- Different scenario parts can also spatially *succeed* each other, with new parts evolving from final positions & results produced by the previous parts.
- This potentially parallel and distributed scenario evolution may proceed in *synchronous* or *asynchronous* mode, also their any combinations.
- SGL operations and decisions in evolving scenario parts can use control states and values *returned* from other scenario parts whatever complex and remote they might be, thus combining *forward* and *backward* scenario evolution in distributed spaces.
- Different steps from the same or different scenario parts can be temporarily associated with the same, reached, world points, sharing persistent or provisional information in them.
- Staying with world points, it is possible to *change* local parameters in them, whether physical or virtual, thus *impacting* the navigated worlds via these locations.
- Scenarios navigating distributed spaces can *create arbitrary distributed physical or virtual infrastructures* in them, which may operate on their own after becoming active, with or without additional external control. They can also be subsequently (or even during their creation) navigated, updated, and processed by same or other scenarios.
- Overall organization of the world creation, navigation, coverage, modification, analysis, and processing can be provided by a *variety of SGL rules* which may be arbitrarily *nested*.
- The evolving SGL scenario, as already mentioned, can *lose utilized parts* if not needed any more; it can also *self-modify* and *self-replicate* during space navigation, to adjust to unknown environments and reduce communications in distributed systems.

3.4.3   Sense and Nature of SGL Rules

Some more light on the general sense and nature of rules which, capable of representing in SGL any actions or decisions, may belong to the following categories:

- Elementary arithmetic, string, or logic operation.
- Move or hop in a physical, virtual, execution, or combined space.
- Hierarchical fusion and return of (potentially remote) data.
- Distributed control, sequential and/or parallel, in both breadth and depth of the scenario evolution.
- A variety of special contexts detailing navigation in space, also character and peculiarities of the embraced operations and decisions.
- Type and sense of a value or its chosen usage, guiding automatic language interpretation.
- Individual or massive creation or removal of nodes and connecting links in distributed knowledge networks, allowing us to effectively work with arbitrary knowledge structures, including their initial establishment and any modification.
- A rule can also be a compound one integrating other rules; it can also be a result of another scenario of any complexity and world coverage.

All rules, regardless of their nature, sense or complexity, are pursuing the same unified ideology and organization, as follows.

- They start from a certain world position, being initially linked to it.
- Perform or control the needed operations in a distributed space, which may be stepwise, parallel and arbitrarily complex, both local and remote.
- Produce or supervise concluding results of the scenario embraced, expressed by control states and values.
- These results may associate with the same (where the rule started) or other world positions, reached by the rule's activity.

This uniformity allows us to effectively compose integral and transparent spatial algorithms of *any complexity and world coverage*, operating altogether under unified and automatic (generally parallel and distributed) control.

3.4.4   The Use of SGL Variables

Let us consider some more details on the nature and sense of spatial variables, stationary or mobile, which can be used in fully distributed physical, virtual or executive environments, effectively serving multiple cooperative processes under the unified control. They are created upon declaration by special rules or by first assignment to them.

- *Global variables*—the most expensive ones, which can serve any SGL scenarios and can be shared by their different branches. Their locations, mobility capabilities, and life span can depend on the features of distributed environments and SGL implementations.

- *Heritable variables*—stationary, appearing within a scenario step and serving only all subsequent steps, generally multiple and parallel (but not from other branches), which can share them in both read and write operations.
- *Frontal variables*—mobile, temporarily associated with the current step and not shared with other parallel steps; they are accompanying scenario evolution, being transferred between subsequent steps. These variables replicate if from a step a number of other steps emerge directly. (The replication procedure, also physical mobility, may have implementation peculiarities if working with physical matter rather than information as frontal variable contents.)
- *Environmental variables*—these allow us to access, analyze, and possibly change different features of physical, virtual and executive words during their navigation, also key parameters of the underlying language implementation system. Most of them are stationary, associated with the world positions reached, but some, especially related to the details of the language interpretation, can be mobile, some even global like the absolute time.
- *Nodal variables*—stationary, being a sole property of the world positions reached by the scenarios. Staying at world nodes, they can be accessed and shared by all activities having reached these nodes under the same scenario identity, and at any time.

These types of variables, especially when used together, allow us to create advanced algorithms working *directly in space*, actually *in between* components of distributed systems rather than *in* them, providing flexible, robust and self-recovering solutions (stealthy too, if needed). Such algorithms can freely self-replicate, partition, spread and migrate in distributed environments (*partially or as an organized whole*), while always preserving overall awareness and global goal orientation.

3.4.5   SGL Control States and Their Hierarchical Merge

The following control states can appear after completion of different scenario steps. Indicating local progress or failure they can be used for effective control of multiple distributed processes with proper decisions at a variety of levels.

- thru—reflects *full success* of the current scenario branch with capability of further development (i.e. indicating successful operation not only *in* but also *through* this step of control). The following scenario steps, if any, will be allowed to proceed from the final location reached by the current step.
- done—indicates success of the current scenario step with its *planned termination*, after which no further development of this branch from the current step and location reached will be possible. This state can, however, be subsequently changed to *thru* at higher levels by a special rule.

- `fail`—indicates non-revocable failure of the current branch, with no possibility of further development from the location reached. This state directly relates to the current branch and step only, but can influence decisions at higher levels by special rules supervising engagement of other branches too (same can be said about the previous two states).
- `fatal`—reports *fatal, terminal failure* with nonlocal effect, triggering massive abortion of all currently evolving scenario processes and removal of the associated temporary data with them, regardless of their current locations and operational success. The scope of this spreading termination may be the whole scenario, by default, or may be restricted by a special containment rule supervising the scenario part within which this state can potentially occur.

These control states appearing in different branches of parallel and distributed scenario at bottom levels can be used to obtain generalized control states at higher levels, up to the whole scenario, in order to make proper decisions for the scenario evolution. The hierarchical bottom-up merge and generalization of states is based on their comparative importance, or strength, where the stronger state will always dominate while ascending towards the decision root.

For example, merging states `thru` and `done` will result in `thru`, thus generally classifying successful development at a higher scenario level with possibility of further expansion from at least some of its branches. Merging `thru` and `fail` will result in `thru` too, indicating general success with possibility of further development despite some branch (or branches) failed while others remain open to further evolution. Merging `done` and `fail` will result in `done` indicating generally successful termination while ignoring local failures, however, without possibility of further development in all these directions. And `fatal` will always dominate when merging with any other states until its destructive influence is contained within a certain higher level rule, as already mentioned (the latter will itself terminate with *fail* in such a case). So ordering these four states by their powers from maximum to minimum will be as follows: `fatal`, `thru`, `done`, `fail`.

These four states, their merge procedure and the use in control rules are standard, language-embedded features. SGL, as a universal spatial language, also allows us to artificially set up any imaginable control states, with any values and numbers, also any merge or generalization procedures, which may include the mentioned standard ones or be completely different.

3.5 Elementary Examples in SGL

Let us consider only some elementary scenario examples in SGL from the mentioned three worlds (PW, VW, and EW). They all will be expressed within the same unified SGL syntax of Fig. 3.5, with full language constructs and details described in the following chapter.

Depending on the concrete language implementation, some flexibilities and deviations from this universal rule-based syntax could be permissible, like using traditional symbols of operations and delimiters, also omitting redundant clarifications without which the scenario text can be understood and interpreted correctly too. This will be mentioned together with the examples exhibited for reducing and simplifying the scenario texts, with overall scenario compositions remaining, however, within the overall organization of Fig. 3.5.

Example 1 Assignment of the sum of three values 27, 33 and 55.6 to a variable named `Result`, as in Fig. 3.6:

```
assign(Result, add(27, 33, 55.6))
```

The variable `Result` will be created, if not existing yet, and will be associated together with the obtained value with the world position where the scenario started. The addition and assignment operations will be accomplished, correspondingly, by rules `add` and `assign`, with the resultant control to remain in the current world position (virtual, physical, executive, or combined) too. Simplified and shorted version with the mentioned above deviations may be as follows:

```
Result = 27 + 33 + 55.6
```

Example 2 Move from the current location in physical space into other location with coordinates (x1, y3), as in Fig. 3.7:

```
move(location(x1, y3))
```

This will cause physical movement from the current world position to the new physical position with given coordinates, with resultant control moving to the reached destination too. The rule `location` clarifies the meaning of its parameters x1 and y3 as coordinates, and rule `move` performs physical movement to the

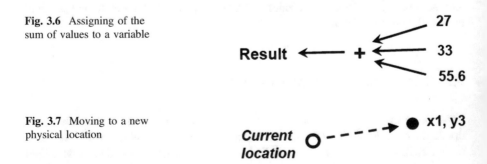

Fig. 3.6 Assigning of the sum of values to a variable

Fig. 3.7 Moving to a new physical location

Fig. 3.8 Moving
independently to two physical
locations

location needed. The shortened version may be as follows (if rule move is always
used with physical coordinates of the destination needed):

```
move(x1, y3)
```

Example 3 Move physically from the current location independently and simultaneously to locations (x1, y3) and (x5, y8), see Fig. 3.8:

```
branch(move(location(x1, y3),
    move(location(x5, y8)))
```

Will cause movement from the current physical position to the two new physical
positions by given coordinates independently and possibly in parallel (if the latter
supported by implementation) using additional rule branch, with the resultant
control split and moving to both destinations too. Simplified version may be as
follows if independent branches can be simply separated by a comma and rules
branch and location omitted:

```
move(x1, y3), move(x5, y8)
```

Further simplification can be by using comma for separation of coordinates of
the two independent destinations under the common rule move as follows:

```
move((x1, y3), (x5, y8))
```

Example 4 Move from a current location first to location (x1, y3) and then to
location (x5, y8) from the same starting location, see Fig. 3.9:

```
sequence(move(location(x1, y3),
    move(location(x5, y8)))
```

Fig. 3.9 Moving
sequentially to two physical
locations

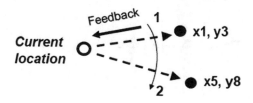

Fig. 3.10 Successive
movement in physical space

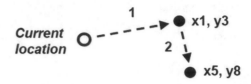

Similar to the previous case but movement to the second location from the same starting position will take place only after full completion of the movement to the first location (using rule sequence for launching two branches both starting from the same world position). The resultant control will split and remain in the two locations reached, as in the previous case. Simplified and reduced version may be as follows:

```
sequence_move((x1, y3), (x5, y8))
```

Example 5 Move from the current location to location (x1, y3) and then from it to location (x5, y8), as in Fig. 3.10:

```
advance(move(location(x1, y3),
    move(location(x5, y8)))
```

The movement will first take place to location (x1, y3) and then will continue from it to location (x5, y8), with the resultant control staying only in the second location upon reaching it (using rule advance for successive movement in space). Shortened version using semicolon for separation of successive operations, each starting from results and place of the previous operation, and omitting the rules advance and location will be as follows:

```
move(x1, y3); move(x5, y8)    or
move((x1, y3); (x5, y8))
```

Example 6 Creation of a virtual node Peter, see Fig. 3.11:

```
create(node(Peter))
```

Starting from the current world location, a new, isolated, virtual node with the given name will be created with the resultant control moving into it. The rule node explains the sense of the item enclosed, and rule create activates the node

Fig. 3.11 Creating a virtual
node

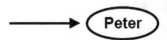

Fig. 3.12 Network extension
with link-node pair

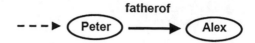

creation procedure. Shortened version may be after omission of the rule `node`, as a
single parameter in rule `create` can only stay for a node name, as follows:

```
create(Peter)
```

Example 7 Extending the virtual network (already having node `Peter`) with a
new link-node pair stating that "Peter is father of Alex", see Fig. 3.12:

```
advance(hop(node(Peter)),
    create(link(+fatherof), node(Alex)))
```

The scenario first directly hops into the already existing node "Peter" and from it
creates new link-node pair with both link and node properly named, where the
succession in virtual space is provided by rule `advance`, similar to how it acted in
physical space in one of previous examples. The rule `link` similar to `node`
clarifies the sense of the embraced item (its sign setting orientation), and the
resulting control will reside in the new node `Alex`.

Simplified version will be as follows, with omission of rules `link` and `node`, as
a sequence of two parameters in rule `create` always treats the first one as link
name and the second as node name:

```
hop(Peter); create(+fatherof, Alex)
```

Example 8 Creating the whole network from scratch, see Fig. 3.13:

```
advance(create(node(Peter)),
    create(link(+fatherof), node(Alex)))
```

Creates the whole network consisting of two nodes and oriented link between
them, with control first moving into node `Peter` and then via the link
`+fatherof` to node `Alex`, as before. Simplified version for this case may be as:

```
create(Peter); create(+fatherof; Alex) or
create(Peter; (+fatherof; Alex))
```

Fig. 3.13 Creating the whole
network

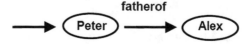

Fig. 3.14 Ordering a robot to
shoot by coordinates

Example 9 Giving direct command to robot `Shooter` to fire at coordinates
(x, y), as in Fig. 3.14:

```
advance(hop(robot(Shooter)), fire(location(x, y)))
```

The scenario moves from a starting position into robot `Shooter` and then
activates the shooting operation in it by the coordinates given, with the resultant
control remaining in the robot. Simplified version may be if name `Shooter`
usually associates with a robot and not a soldier, and rule `fire` normally uses the
target coordinates as parameters, so additional explanations may be redundant, as
follows:

```
hop(Shooter); fire(x, y)
```

Example 10 Order soldier `John` to use robot `Shooter` to fire at coordinates
(x, y) with confirmation of the robot's success or failure by the soldier, see
Fig. 3.15:

```
advance(
  hop(soldier(John)),
  if(advance(
    hop(robot(Shooter)), fire(location(x, y))),
    report(success),
    report(failure)))
```

The scenario first moves to soldier `John` and then passes to robot `Shooter`
ordering the latter to perform the firing operation by coordinates given. The soldier
waits for the result of robot's activation & shooting and on its completion reports to

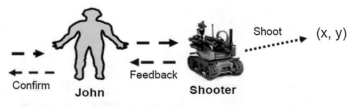

Fig. 3.15 Ordering soldier to use robot to shoot by coordinates, confirming the result

a higher level (rule `report`) about its success or failure, with the resulting control staying with the soldier. The three branches of rule `if` will coordinate this, where by success or failure of the first branch either second or third branch will continue working.

Simplified version will be as follows (if names `John` and `Shooter` used to be associated, correspondingly, with the soldier and robot and additional explanations not needed, also semicolon used instead of rule `advance`):

```
hop(John); if((hop(Shooter); fire(x, y)),
            report(success),
            report(failure))
```

Another variant of this scenario with rule `report` shifted to a higher level can be as follows:

```
hop(John); report(
  if((hop(Shooter); fire(x, y)), success, failure))
```

Example 11 Find and output in the current position the maximum temperature measured in places with coordinates (x1, y1), (x2, y2), and (x3, y3) visited independently from the same starting location, see Fig. 3.16, with Ti symbolizing temperatures in the locations.

```
output(
   max(
    advance(
     branch(
      move(x1, y1), move(x2, y2), move(x3, y3)),
      TEMPERATURE)))
```

Fig. 3.16 Finding and reporting the maximum temperature among remote locations

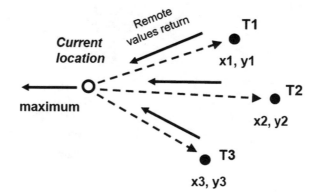

The scenario first moves independently (possibly, in parallel if technically feasible, using rule `branch`) to three physical locations with given coordinates and then measures temperature in each of them (using special environmental variable `TEMPERATURE` in the locations after reaching them, with rule `advance` providing this succession). The obtained values (which may be arbitrarily remote) are returned to the starting position with selecting the largest one (using rule `max`) and sending it (rule `output`) to a terminal associated with the starting position, with the resultant control staying in it too. Simplified version will be as follows:

```
output(max(move((x1, y1), (x2, y2), (x3, y3));
      TEMPERATURE))
```

Or, with reduced parentheses using aggregate rule `output_max` over the same operand:

```
output_max(move((x1, y1), (x2, y2), (x3, y3));
      TEMPERATURE)
```

Example 12 Find the maximum temperature measured in places with coordinates (x1, y1), (x2, y2), and (x3, y3) visiting them in a strict succession and outputting the results in the final place, as in Fig. 3.17:

```
advance(
   frontal(Points, Maximum),
   assign(Points, ((x1, y1), (x2, y2), (x3, y3)))),
 repeat(
  advance(
    nonempty(Points),
```

Fig. 3.17 Finding maximum temperature in a consecutive order

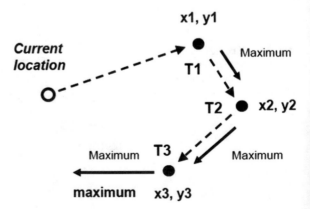

```
   move(withdraw(Points, first)),
  assign(Maximum, max(Maximum, TEMPERATURE))))),
  output(Maximum))
```

The scenario, first declaring two variables Points (to keep and transfer coordinates of the points to be visited) and Maximum (to update and transfer the maximum temperature value) as frontal, moving ones, visits the three locations in a succession, one after the other, each time obtaining next location coordinates from the head of their list (together with removing them from the list) using rule withdraw. In each of the points reached it updates the maximum value in variable Maximum transferred between the locations, comparing it with the temperature in the visited location, thus receiving the absolute maximum on the three in final location.

It then outputs this maximum at a terminal associated with the final location using rule output, also leaving control there. The rule repeat provides repetitive succession of the movements in space, which will stop in the final position (where frontal variable Points becomes empty). This scenario will also work for any number of world positions to be visited in a succession, with their coordinates to be placed as a list into variable Points at the beginning. Simplified & shortened version may be as follows:

```
frontal(Points = (x1_y1, x2_y2, x3_y3), Maximum);
repeat(move(withdraw(Points, first));
    Maximum = max(Maximum, TEMPERATURE));
  output(Maximum)
```

Where underscore is used to aggregate two dimensions of coordinates for reducing parentheses, and rule frontal has a broader scope, embracing assignment to Points. The verification rule nonempty is omitted as the following rule move will inevitably stop the repeat construct after rule withdraw finds the content of frontal variable Points empty, thus being unable to perform the repetition further. After this, the next rule output will work with the final content in frontal variable Maximum.

We have shown here only elementary examples in SGL, just to have an initial look at the language capabilities and usage, with numerous programming examples presented and explained later throughout the book. The detailed language description with explanation of syntax and semantics of its constructs will be given in the next chapter.

3.6 General Issues of SGL Networked Interpretation

The SGM if used in distributed environments operates as follows. A network of
SGL interpreters (as universal control modules U, Fig. 3.18) embedded in key
system points (humans, robots, sensors, smart phones, smart watches, etc.)
throughout any space of interest collectively interprets high-level mission scenarios
written in SGL. Capable of representing any parallel and distributed algorithms
these scenarios can start from any node (or nodes), runtime covering the whole
world or its parts needed with operations and control via the interpretation network.

The self-spreading scenarios can create knowledge infrastructures, both passive
and active, arbitrarily distributed between system components, as in Fig. 3.19.
Navigated by same or other scenarios, these can effectively support distributed
databases, command and control (C2), situation awareness, and autonomous
decisions, also simulate any other existing or hypothetic computational and/or
control models.

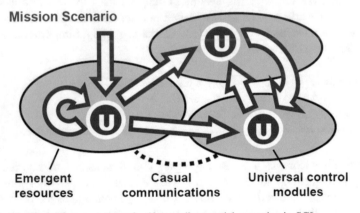

Fig. 3.18 Distributed interpretation of self-spreading spatial scenarios in SGL

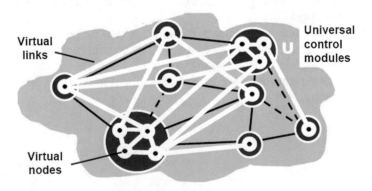

Fig. 3.19 Creation of spatial infrastructures

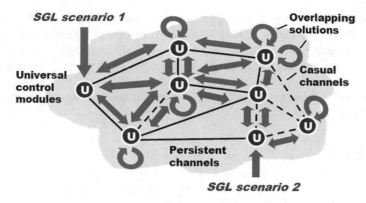

Fig. 3.20 Spatial interaction of different scenarios

Many SGL scenarios can simultaneously operate within the same distributed environments, spatially cooperating or competing in the networked space as overlapping fields of solutions, see Fig. 3.20.

The dynamic network of SGL interpreters covering distributed spaces may have any (including runtime changing) topology and can operate without any central facilities or control, exhibiting at the same time wholeness and high integrity as a system. As will be shown later, the overall management of distributed evolution of high-level SGL scenarios can be based on a special *track infrastructure* supporting overall awareness, goal orientation, automatic C2, also properly handling various distributed information resources (including their timely creation, optimization, and cleaning/removal). The distributed execution of SGL scenarios can be effectively implemented in a variety of systems, whether technical or social, with any types of communications between their components (electronic, acoustic, visual, postal, even direct voice or paper writing).

The technology can also convert any purely human collectives into holistic systems operating under local and global goals and capable of acting even without any special electronic equipment, after being taught of how to interpret, modify, partition, and distribute the received scenarios, also how to maintain contacts and subordination with other individuals within the collective scenario interpretation.

Chapter 6 will be describing organization of the networked SGL interpreter in detail.

3.7 Conclusion

We have considered the main features of Spatial Grasp Model (SGM) and its key element Spatial Grasp Language (SGL) for creation, analysis, and management of large distributed dynamic systems. The approach is based on self-evolving and self-propagating patterns in SGL which are spatially matching distributed environments in parallel wavelike mode. The waves in SGM are considered as active substances bringing activity in the form of operations and control into the systems navigated and covered, contrary to usual understanding of waves as a result of system's activity, not a source (like, for example, brain waves). This can provide enormous flexibility, power, and opportunity to create and manage large systems as a whole, with numerous applications considered in the subsequent chapters.

References

1. R. Gray, D. Kotz, S. Nog, D. Rus, G. Cybenko, Mobile agents for mobile computing. Technical Report PCS-TR96-285, 2 May 1996.
2. Rodič, B. (2011). Mobile agents for distributed decision support systems. *Manag. Inf. Syst.*, 6(1), 020–027.
3. P.S. Sapaty, A wave approach to the languages for semantic networks processing, in *Proceedings of International Workshop on Knowledge Representation*. Section 1: Artificial Intelligence, Kiev (1984, in Russian).
4. P.S. Sapaty, A wave language for parallel processing of semantic networks. Comput. Artif. Intell. 5(4) (1986).
5. S. Varbanov, P.S. Sapaty, An information system based on the wave navigation techniques, in *Abstracts of the International Conference on AIMSA '86*, Varna, Bulgaria (1986)
6. P.S. Sapaty, The WAVE-0 language as a framework of navigational structures for knowledge bases using semantic networks, in *Proceedings of USSR Academy of Sciences*. Technical Cybernetics, No. 5 (1986, in Russian).
7. P.S. Sapaty, I. Kocis, A parallel network wave machine, in *Proceedings of the 3rd International Workshop PARCELLA '86* (Akademie-Verlag, Berlin, 1986)
8. P.S. Sapaty, The wave approach to distributed processing of graphs and networks, in *Proceedings of the International Working Conference on Knowledge and Vision Processing Systems*, Smolenice, Nov 1986
9. P.S. Sapaty, S. Varbanov, M. Dimitrova, Information systems based on the wave navigation techniques and their implementation on parallel computers, in *Proceedings of the International Working Conference on Knowledge and Vision Processing Systems*, Smolenice, Nov 1986
10. P. Sapaty, S. Varbanov, A. Iljenko, The WAVE model and architecture for knowledge processing, in *Proceedings of Fourth International Conference on Artificial Intelligence and Information-Control Systems of Robots*, Smolenice, 1987
11. P.S. Sapaty, The WAVE-1: a new ideology and language of distributed processing on graphs and networks. Comput. Artif. Intell. No.5 (1987)
12. P.S. Sapaty, WAVE-1: a new ideology of parallel processing on graphs and networks, in *Proceedings of International Conference on Frontiers in Computing*, Amsterdam (1987)
13. P.S. Sapaty, WAVE-1: a new ideology of parallel processing on graphs and networks, in *Future Generations Computer Systems*, vol. 4, North-Holland (1988)
14. P.S. Sapaty, The WAVE model for advanced knowledge processing. Report No. OUEL 1803/89, University of Oxford, England (1989)
15. P.S. Sapaty, The WAVE Model for advanced knowledge processing, in *CAD Accelerators*, ed. by A.P. Ambler, P. Agrawal, W.R. Moore (Elsevier Science Publ. B.V., 1990)
16. P.S. Sapaty, The WAVE machine project, in *Proceedings of the IFIP Workshop on Silicon Architectures for Neural Nets*, St. Paul de Vence, France, 28–30 Nov 1990
17. P.S. Sapaty, Logic flow in active data, in *VLSI for Artificial Intelligence and Neural Networks*, ed. by W.R. Moore, J. Delgado-Frias (Plenum Press, New York, 1991)

18. P.S. Sapaty, W. Zorn, The WAVE model for parallel processing and its application to computer network management, in *International Networking Conference INET'91*, Copenhagen, 1991

19. P.S. Sapaty, The WAVE paradigm, in *Proceedings of the JICSLP'92 Post-Conference Joint Workshop on Distributed and Parallel Implementations of Logic Programming Systems*, Washington, D.C., 13–14 Nov 1992

20. L. Bic., P. Borst, M. Corbin, P. Sapaty, The WAVE control protocol for distributed interactive simulation, in *Proceedings of the 11th International Conference on Interoperability of Distributed Simulations, IST UCF*, Orlando, FL, 26–30 Sept 1994

21. P.M. Borst, M.J. Corbin, P.S. Sapaty, WAVE processing of networks and distributed simulation, in *Proceedings of the HPDC-3 International Conference*, San Francisco, Aug 1994 (IEEE), pp. 61–69

22. M. Corbin, P.S. Sapaty, Using the WAVE paradigm for parallel simulation in distributed systems, in *Proceedings of the International Conference ParCo93*, Grenoble, France, Sept 1993. Also in *Parallel Computing: Trends and Applications*, ed. by G.R. Joubert, D. Trystram, F.J. Peters, D.J. Evans (North-Holland, 1994)

23. P.S. Sapaty, M. Corbin, P.M. Borst, Using the WAVE paradigm for modeling and control of dynamic multi-agent systems, in *Poster at the Artificial Life IV Conference*, 6–8 July 1994, Massachusetts Institute of Technology, Cambridge, MA

24. P.S. Sapaty, P.M. Borst, An overview of the WAVE language and system for distributed processing in open networks. Technical Report, Department of Electronic & Electrical Engineering, University of Surrey, June 1994

25. P.S. Sapaty, M. Corbin, P.M. Borst, A. Went, WAVE: a new technology for intelligent control in communication networks, in *Proceedings of the International Conference on "The Application of RF, Microwave and Millimetre Wave Technologies" (M'94)*, Wembley, UK, 25–27 Oct 1994, Nexus, pp. 434–438

26. P.S. Sapaty, M.J. Corbin, P.M. Borst, Mobile WAVE programming as a basis for distributed simulation and control of dynamic open systems. Report at the 4th UK SIWG National Meeting, SGI Reality Centre, Theale, Reading, 11 Oct 1994

27. P.S. Sapaty, M.J. Corbin, P.M. Borst, Mobile WAVE programming as a basis for distributed simulation and control of dynamic open systems, in *A Special Session on the WAVE Technology at the 15th International Conference on Distributed Computing Systems*, May–June 1995, Vancouver, BC, Canada

28. M.J. Corbin, P.S. Sapaty, Distributed object-based simulation in WAVE. J. Simul. Pract. Theory **3**(3), 157–181 (1995)

29. F. Merchant, L.F. Bic, P.M. Borst, M.J. Corbin, M. Dillencourt, M. Fukuda, P.S. Sapaty, Simulating autonomous objects in a spatial database using WAVE, in *Proceedings of the 9th European Simulation Multiconference*, Prague, June 1995

30. P.S. Sapaty, Mobile wave technology for distributed knowledge processing in open networks, in *Proceedings of the Workshop on New Paradigms in Information Visualization and Manipulation, in conjunction with the Fourth International Conference on Information and Knowledge Management (CIKM'95)*, Baltimore, Maryland, Dec 1995

31. P.S. Sapaty, P.M. Borst, WAVE: mobile intelligence in open networks, in *Proceedings of the First Annual Conference on Emerging Technologies and Applications in Communications (etaCOM)*, Portland, Oregon, May 1996 (IEEE Computer Society Press, 1996), pp. 192–195

32. P.S. Sapaty, WAVE: creating dynamic worlds based on mobile cooperative agents, Dartmouth Workshop on Transportable Agents, Dartmouth College, Hanover, New Hampshire, Sept 1996

33. J.C.C. Darling, P.S. Sapaty, Distributed dynamic virtual reality in WAVE, in *Proceedings of the European Simulation Symposium (ESS-96)*, Genoa, Italy, Oct 1996, pp. 36–40

34. P.S. Sapaty, Live demonstration of the WAVE system and applications at the Workshop on Mobile Agents and Security 97, Maryland Center for Telecommunications Research, Department of Computer Science and Electrical Engineering, UMBC, 27–28 Oct 1997

35. P.S. Sapaty, Mobile Programming in WAVE, in *Mathematical Machines and Systems*, No. 1, January–March 1998, Kiev, pp. 3–31, ISSN: 1028–9763
36. P.S. Sapaty, *Mobile Processing in Distributed and Open Environments* (Wiley, New York, 1999), 436 p, ISBN: 0471195723
37. P.S. Sapaty, Cooperative conquest of distributed worlds in WAVE, in *Proceedings of the Symposium and Exhibition of the Unmanned Systems of the New Millennium, AUVSI'99*, Baltimore, MD, 13–15 July 1999, 16 p
38. Sapaty, P. S. (2000). Cooperative exploration of distributed worlds in WAVE. *International Journal of Artificial Life and Robotics, 4,* 109–118.
39. P.S. Sapaty, High-level spatial scenarios in WAVE, in *Proceedings of the International Symposium AROB 5th*, Oita, Japan, Jan 2000, pp. 301–304
40. P.S. Sapaty, Spatial Programming of Distributed Dynamic Worlds in WAVE, in *Presentation at the Special Colloquium "Internet Challenges"*, Hasso-Plattner-Institut, University of Potsdam, Berlin, Germany, 4 Oct 2002, 50 p
41. P. Sapaty, M. Sugisaka, WAVE-WP (World Processing) Technology, in *Proceedings of the First International Conference on Informatics in Control, Automation and Robotics,* vol 1, Setubal, Portugal, 25–28 Aug 2004, pp. 92–102
42. P.S. Sapaty, WAVE-WP (World Processing) Technology. Math. Mach. Syst., No. 3 (2004), pp. 3–17, ISSN: 1028-9763
43. Sapaty, P. (2005). *Ruling Distributed Dynamic Worlds*. New York: Wiley.
44. P. Sapaty, High-level technology to manage distributed robotized systems, in *Proceedings of the Military Robotics 2010*, 25–27 May, Jolly St Ermins, London UK
45. P. Sapaty, Providing global awareness in distributed dynamic environments, in *International Summit ISR*, London, 16–18 Apr 2013.
46. P. Sapaty, The world as an integral distributed brain under spatial grasp paradigm, in *Intelligent Systems for Science and Information* (Springer, New York, 2014)
47. P. Sapaty, Towards massively robotized systems under spatial grasp technology. J. Comput. Sci. Syst. Biol. **9**(1) (2016)
48. P. Sapaty, A brief introduction to the spatial grasp language (SGL). J. Comput. Sci. Syst. Biol. **9**(2) (2016)
49. P.S. Sapaty, Towards global goal orientation, robustness and integrity of distributed dynamic systems. J. Int. Relat. Diplomacy **4**(6) (2016)

Chapter 4
SGL Detailed Specification

4.1 Introduction

SGL [1–4] is the latest and most advanced version in a sequence of spatial languages based on free and cooperative propagation of active waves in networks, with the previous versions named as WAVE [5–11], WAVE-WP (World Processing) [12], WPL (World Processing Language) [13], and DSL (Distributed Scenario Language) [14].

The chapter contains full specification of improved and extended SGL (with its programs-scenarios called *grasps*) providing detailed syntax and semantics of the main language constructs: *constants*, different types of *spatial variables*, and powerful and universal operational and descriptive units called *rules*. Due to recursion, both constants and rules can themselves be arbitrary *grasps*, resulting altogether in flexible spatial scenarios at different system levels and their mixtures. This allows SGL to be used as a single, and only, language for integral holistic system management without traditional parts and layers at different languages and time and space consuming interfaces in between.

SGL effectively supports parallel forward and echo processes, generalization of distributed situations and direct processing of remote knowledge, with obtained results retained in the navigated environments or returned to the starting points. The universal recursive SGL structure also allows the language to be easily modified and extended for any new classes of distributed applications.

4.2 Full SGL Syntax and Main Constructs

The following is full SGL description extending and detailing the mentioned in previous chapter main language constructs and features. Syntactic categories are shown in italics, vertical bar separates alternatives, parts in braces indicate zero or

© Springer International Publishing AG 2017
P.S. Sapaty, *Managing Distributed Dynamic Systems with Spatial Grasp Technology*, Studies in Computational Intelligence 690,
DOI 10.1007/978-3-319-50461-2_4

more repetitions with a shown delimiter at the right if more than one, and constructs in brackets are optional. The remaining characters and words are the language symbols (including boldfaced braces).

grasp → *constant* | *variable* | *rule* [({ *grasp*, })]

constant → *information* | *matter* | *custom* | *special* | *grasp*

variable → *global* | *heritable* | *frontal* | *nodal* | *environmental*

rule → *movement* | *creation* | *echoing* | *verification* | *assignment* | *advancement* | *branching* | *transference* | *timing* | *granting* | *type* | *usage* | *application* | *grasp*

information → *string* | *scenario* | *number*

string → ` {*character*} '

scenario → { {*character*} }

number → [*sign*]{*digit*}[. {*digit*}[e[*sign*]{*digit*}]]

matter → " {*character*} "

special → thru | done | fail | fatal | infinite | nil | any | all | other | passed | existing | neighbors | direct | noback | around | firstcome | forward | backward | global | local | along | against | neutral | synchronous | asynchronous | virtual | physical | executive | engaged | vacant

global → G{*alphameric*}

heritable → H{*alphameric*}

frontal → F{*alphameric*}

nodal → N{*alphameric*}

environmental → TYPE | CONTENT | NAME | ADDRESS | QUALITIES | WHERE | BACK | PREVIOUS | PREDECESSOR | DOER | RESOURCES | LINK | DIRECTION | WHEN | TIME | SPEED | STATE | VALUE | COLOR | IN | OUT | STATUS | TEMPERATURE | PRESSURE | HUMIDITY | VISIBILITY | ELEVATION

movement	→	`hop｜move｜shift｜follow｜destination`
creation	→	`create｜linkup｜delete｜unlink`
echoing	→	`state｜order｜rake｜sum｜count｜first｜`
		`last｜min｜max｜random｜average｜`
		`quickest｜element｜sortup｜sortdown｜`
		`reverse｜add｜subtract｜multiply｜divide｜`
		`degree｜separate｜unite｜attach｜append｜`
		`common｜withdraw｜access｜invert`
verification	→	`equal｜nonequal｜less｜lessorequal｜`
		`more｜moreorequal｜bigger｜smaller｜`
		`heavier｜lighter｜longer｜shorter｜empty｜`
		`nonempty｜belong｜notbelong｜`
		`intersects｜nonintersects`
assignment	→	`assign｜assignpeers`
advancement	→	`advance｜slide｜repeat｜align`
branching	→	`branch｜sequential｜parallel｜if｜or｜`
		`orsequential｜orparallel｜and｜`
		`andsequential｜andparallel｜choose｜`
		`firstrespond｜cycle｜loop｜sling｜`
		`whirl｜split｜fringe`
transference	→	`run｜call｜input｜output｜transmit｜send｜`
		`receive`
timing	→	`sleep｜allowed`
granting	→	`contain｜release｜free｜blind｜stop｜`
		`lift｜none｜stay｜seize`
type	→	`global｜heritable｜frontal｜nodal｜`
		`environmental｜matter｜number｜string｜`
		`scenario`
usage	→	`address｜coordinate｜content｜index｜`
		`time｜speed｜name｜place｜center｜range｜`
		`doer｜human｜soldier｜robot｜nodes｜`
		`links｜unit`

4.3 SGL Constants

4.3.1 Information

String can be represented as any sequence of characters embraced by opening-closing single quotation marks. This sequence should not contain other single quotes inside unless they appear in opening-closing pairs, with any nesting allowed. Examples: 'John', 'Peter and Paul'. If single words representing information are not intersecting with other language constructs, the quotes around them can be omitted: John, Peter, Paul.

Instead of single quotes, a sequence of characters can also be placed into opening-closing curly brackets (or braces, { }), which can be used inside the string in pairs too. Braces will indicate the text as a potential *scenario* code which can be immediately optimized (like removing unnecessary spaces and/or adjusting to the standard SGL syntax, say, after using constructs typical to other programming languages for convenience, as mentioned in the previous chapter). If single quotes are used to embrace texts as a potential SGL code, such code optimization will have to be done *during* its interpretation, not before, and each time it is involved, with the original text remaining intact.

Number can be represented in a standard way, similar to traditional programming languages, generally in the form: $[sign]\{digit\}[.\{digit\}[E[sign]\{digit\}]]$ (with brackets identifying optional and braces repeating characters).

Examples: 105, 88.56, -15, 3.3E-5.

Numbers can also use words instead of digits and accompanying characters: zero, one, two, three, four, five, six, seven, eight, nine, ten, eleven, twelve, thirteen, fourteen, fifteen, sixteen, seventeen, eighteen, nineteen, twenty, thirty, forty, fifty, sixty, seventy, eighty, ninety, hundred, thousand, million, billion, trillion, dot, minus, plus (with underscore as separator if more then one word needed to represent them). The four examples above may also look like:

(a) with mixed representation: hundred_five, eighty_eight.56, minus_fifteen, three.3E-five;
(b) up to the full wording: one_zero_five, eight_eight_dot_fifty_six, minus_one_five, three_dot_three_E_minus_five.

4.3.2 Physical Matter

Matter (incl. physical objects) can be represented by a sequence of characters embraced by opening-closing double quotation marks.

Examples: "truck", "white sand", "brick", "water".

The above mentioned self-identified constants (i.e. strings, scenarios, numbers, and matter) may also be set up by explicit naming their types with the use of corresponding rules.

4.3.3 Custom Constants

For extended applications, other self-identified or *custom* constants can also be introduced, if not conflict with the language syntax, to be directly interpreted by an extended SGL interpreter. For example, these may be coordinates in physical spaces similar to x17.5, y44.2, z-77, as well as their combination: x17.5_y44.2_z-77, or internet addresses like, for example, www.amazon.com. Special type-defining rules can be used for custom constants too.

4.3.4 Special Constants

Special verbal constants can be used as standard parameters (or modifiers) in different language rules as will be shown later. The basic list of such words (consisting of lower case letters only) with comments on their possible use is as follows:

thru—indicates (or sets) control state as a success with possibility of further evolution.

done—indicates (or sets) control state as a successful termination, with blocking further development.

fail—indicates (or sets) control state as failure, without further development.

fatal—indicates (or sets) control state as absolute failure, with massive abortion of active distributed processes.

infinite—indicates infinitely large value.

nil—indicates no value at all.

any, all, other—stating that any, all, or other (i.e. except the current one) elements under consideration can be used.

passed—hinting that the world nodes to be considered have already been passed by the current scenario branch.

existing—hinting that world nodes with the given names are already existing and should not be created again (i.e. duplicated).

neighbors—stating that the nodes to be accessed are among direct neighbors of the current node, i.e. within a single hop from it by existing links.

direct—stating that the mentioned nodes should be accessed or created (if not exist) from the current node directly, regardless of possible (non) existence of direct links to them.

noback—not allowing to return to the previously occupied node.

firstcome—allowing to access the next-hop nodes only first time with the given scenario ID.

forward, backward—allowing to move from the current node via existing links along or against their orientations (ignored when dealing with non-oriented links, which can be traversed in both directions).

around—depending in specifics of application, hints that movement through a defined physical or virtual region should be by following its fringes rather than inside of it.

global, local—may indicate the scope of operations or the world access in different rules.

along, against, neutral—used to indicate movement along, opposite, or regardless orientation of links between nodes.

synchronous, asynchronous—a modifier setting synchronous or asynchronous mode of operations induced by different rules.

virtual, physical, executive—indicating or setting the type of a node the scenario is currently dealing with (the node can also be of a combined type).

engaged, vacant—indicating or setting the state of a resource the current scenario is dealing with (like, say, human or robot, or any physical, virtual or combined world node).

4.3.5   Compound Constants, Grasps

Constants can also be arbitrarily complex, as aggregates (possibly hierarchical) from elementary types (not necessarily the same) described above, being supported by the full SGL syntax (i.e. generally as *grasps* again). They can be composed by using either standard rules described later or, if not sufficient, any additional, custom ones oriented on specific application areas.

4.4   SGL Variables

The general sense and use of SGL variables have been mentioned in the previous chapter. Different types of variables can be self-identifiable, i.e. by the way their names are written. Variables can also have any names and types if they are explicitly declared by special rules, explained later.

4.4.1   Global, Heritable, Frontal, and Nodal Variables

In the case of self-identification, names of global, heritable, frontal, and nodal variables should start with capital letters G, H, F or N, respectively, followed by a sequence of alphanumeric characters (letters and digits only).

Examples: Globe, H214b, Frontal5, Nina37.

4.4.2   Environmental Variables

All these variables have specific names written in all capital letters, with brief explanation of their sense and usage following.

TYPE—indicates the type of a node the current step associates with. This variable returns the node's type (i.e. virtual, physical, executive, or their combination as a list with more than one value). It can also change the existing type by assigning to it another value (simple or combined) if needed.

CONTENT—returns content of the current node (only if having virtual or executive dimension, or both), which can be any string of characters (in the simplest case the latter just serving as its name). Assigning to CONTENT allows us to change the existing node's content when staying in it. In a purely physical node CONTENT returns nil (as physical nodes can be identified by their addresses only). If a node is of both virtual and executive nature, this variable deals with the virtual one.

NAME—can be used the same as CONTENT, where information associated with virtual nodes can be treated in both ways. But for certain applications the nodes of virtual networks may happen to be more convenient if keeping names and contents separately.

ADDRESS—returns address of the current virtual node. This is read-only variable as node addresses are set up automatically by the underlying distributed interpretation system during creation of virtual nodes, or by an external system it uses (for example, it can be an internet address of the node).

QUALITIES—identifies a list of available physical parameters associated with the current physical position, or node, depending on the chosen implementation and application (for example, these may be temperature, humidity, air pressure, visibility, radiation, noise or pollution level, density, salinity, etc.). These parameters (generally as a list of values) can be obtained by reading the variable. They may also be changed (depending on their nature and implementation system capabilities) by assigning new values to QUALITIES, thus locally influencing the world from its particular point (or at least attempting to).

WHERE—keeps physical coordinates of the current physical node in the chosen coordinate system (the node can be combined one, additionally having virtual and/or executive features). These coordinates can be obtained by reading the variable. Assigning a new value to this variable can cause physical movement of the current node into the new position (while preserving its identity, all information surrounding, and control and data links with other nodes).

BACK—keeps internal system link to the preceding world node (virtual, executive or combined one with virtual or executive dimension), allowing the scenario to most efficiently return to the previously occupied node, if needed. Referring to internal interpretation mechanisms only, the content of BACK cannot be lifted, recorded, or changed from the scenario level.

PREVIOUS—refers to an absolute and unique address of the previous virtual node (or combined with executive and/or physical dimensions), allowing us to return to the node directly. This may be more expensive than using BACK, but the content of PREVIOUS, unlike BACK, can be lifted, recorded, and used elsewhere in the scenario.

PREDECESSOR—refers to the content/name of the preceding world node (the one with virtual or executive dimension). Its content can be lifted, recorded, and used subsequently, including for organization of direct hops to this node. Assigning to PREDECESSOR can change content/name of the previous node.

DOER—keeps a name of the device (say, laptop, robot, smart sensor, or even a human) which interprets the current SGL code. This device can be chosen for the scenario automatically from the list of recommended devices, or just picked up from those guessed to be available. It can also be appointed explicitly by assigning its name to DOER, causing the current SGL code move immediately into this device and execute there unless it terminates or another device is assigned to DOER, say, when the current one becomes inefficient or fails.

RESOURCES—keeps a list of available or recommended resources (human, robotic, electronic, mechanical, etc., by their types or names) which can be used for execution of the current and subsequent parts of the SGL scenario. This list can contain potential doers too, which after being selected by different scenario branches appear (by their names) in variables DOER associated with the branches. RESOURCES can be accessed and changed by assignment, and in case of distributed SGL interpretation it can be replicated with its content, the latter, possibly, partitioned between different branches by the internal interpretation planning and optimization procedures.

LINK—keeps a name (same as content) of the virtual link which has just been passed. By assigning new value to it you can change the link's content/name. Assigning nil or empty to LINK removes the link passed.

DIRECTION—keeps direction (along, against, or neutral) of the passed virtual link. Assigning to this variable values like plus, minus, or nil (same as +, -, or empty) can change its orientation or make non-oriented.

WHEN—assigning value to this variable sets up an absolute starting time for the following scenario branch, thus allowing us to suspend and schedule certain operations and their groups in time.

TIME—returns current absolute time, being read-only global variable.

SPEED—reflects speed of physical movement of the node (physical, executive or combined, the latter may include virtual dimension too) in which control (represented by the current step) is staying. By assigning to this variable, you can change the speed of the current node. In case of a pure virtual node, the notion of speed is irrelevant and will return nil when accessed, also causing no effect when assigned to.

STATE—can be used for explicit setting of control state of the current step by assigning to it one of the following: thru, done, fail, or fatal. (These states, as mentioned before, are also generated implicitly, automatically on the results of success or failure of different operations, belonging to the overall internal control of scenarios.) Reading STATE will always return thru as this could only be possible if the previous operation terminated with thru too, thus letting this operation to proceed. A certain state explicitly set up in this variable can be used subsequently at higher levels (possibly, together with termination states of other branches) within distributed control provided by SGL rules, whereas assigning fatal to STATE causes already mentioned abortion of multiple distributed processes with the associated data.

VALUE—when accessed, returns the resultant value of the latest operation (say, an assignment to a variable or just naming a variable or constant). Assignment to VALUE leaves its content available to the next operation. This variable allows us to organize balanced processing combining sequences of operations with their representation as nested expressions in SGL. (As follows from the SGL syntax, the resultant values of operations can also be accessed implicitly if these operations or their sequences are themselves standing as operands of higher level rules.)

COLOR—keeps identity of the current SGL scenario or its branch, which propagates together with the scenario and influences grouping of different nodal variables under this identity at world nodes. This means that different scenarios or their branches with different identities are protected from influencing each other via the use of identically named nodal variables. However, scenarios with different colors can penetrate into each other information areas if they know the other's colors, by temporarily assigning the needed new identity to COLOR (to perform cooperative or stealth operations) while restoring the previous color afterwards. Any numerical or string value can be explicitly assigned to COLOR. By default, different scenarios are implicitly assigning the same value in COLOR at the start, thus being capable of sharing all information at navigated nodes, unless change their personal color themselves.

IN—special variable mentioning which in a scenario asks for an additional data from the outside world in the current point of it; this input data becoming its resultant value.

OUT—special variable allowing us to send information from the scenario to the outside world in its current point, by assigning the output value to this variable.

STATUS—retrieving or setting the status of a doer node in which the scenario is currently staying (engaged or vacant, possibly, with a numerical estimate of the level of engagement or vacancy). This feedback from the implementation layer could be useful for a higher-level supervision, planning, and guidance of the use and distribution of resources executing the scenario, rather than doing this fully automatically by standard procedures which may not always be optimal, especially under resource shortages.

TEMPERATURE—returning the current temperature value in the current physical node, being just a single and often used parameter; the whole bunch of such parameters, as a list, being usually represented by the already mentioned variable QUALITIES (reflecting the specifics of concrete SGL implementation). Other such single individual and frequently used parameters may be, for example, PRESSURE, HUMIDITY, VISIBILITY, ELEVATION, etc.

Other environmental variables for extended applications can be introduced and identified by unique words in all capitals too, or they may use any names if explicitly set up by a special rule, as mentioned later. As can be seen, most environmental variables are serving as stationary ones, except RESOURCES and COLOR, which are mobile. The global variable TIME may symbolically be considered as stationary too but in reality may depend on implementation details.

4.5   SGL Rules

The concept of *rule* is not only dominant in SGL for setting most diverse activities ranging from elementary data & knowledge & physical matter processing to overall management and control, but also *the only one*. This provides a universal, integral and unified approach to expressing any operations in distributed dynamic worlds, and if needed, in parallel and fully distributed mode. This section describes the main repertoire of introduced and researched SGL rules with summary of their features.

4.5.1   Movement

Rules of this class result in virtual hopping to the existing nodes (the ones having virtual or executive dimensions) or real movement to new physical locations, associating the remaining scenario (with current frontal variables and control) with the nodes reached. The resultant values of the movements are represented by the reached node names (in case of virtual, executive or combined nodes) or `nil` in case of pure physical nodes, with control state `thru` in them if the movement was successful. If no destinations reached, the movement results with state `fail` and value `nil` in the rule's starting node.

hop—sets virtual propagation to node(s) in virtual, execution, or combined worlds (the latter may have physical dimension too), directly or via semantic links connecting them with the starting node. In case of a direct hop, except node name or address, special modifier `direct` may be included into parameters of the rule. If a hop to take place from a node to a node via an existing link, both destination node name/address and link name (with orientation if needed) should be among parameters of the rule. This hop rule can also cause independent and parallel propagation to a number of nodes if there are more than one node connected to the current one by similarly named links and only link name mentioned (or given by indicator `all`, for all links involved). In a more general case, parallel hops can be organized from the current node if the destination attributes are given by a list of names/addresses of nodes and names of links (or `direct` or `all` indicators) which should lead to them.

move—sets real movement in physical world to a particular location given by coordinates in a chosen coordinate system. The destination location becomes a new temporary node with no name (or `nil`) which disappears when all current scenario activities leave it for other nodes. If, however, the destination node is to have virtual dimension too (indicated by `virtual` in the parameters of the rule, possibly, accompanied by a certain name otherwise default name used), it will remain intact and can be accessed by other scenarios or different branches of the current one until removed explicitly.

shift—differs from the move only in that movement in physical world is set by deviations of physical coordinates from the current position rather than by absolute physical coordinates.

follow—allows us to propagate in both virtual and physical spaces by following arbitrary routes set up by sequences of links, nodes, physical coordinates, etc., or via obtained internal interpretation tracks using recorded entries to them. Some details of using this rule may depend on application areas and implementation details, to be exemplified in some scenarios of subsequent chapters.

destination—sets up certain constraints on the destinations reached by individual movements mentioned above or even being results of arbitrary scenarios, to declare them as valid, which may need nonlocal (say, comparative) operations on the obtained parameters in these nodes or locations.

4.5.2 Creation

This class of rules creates or removes nodes and/or links leading to them during distributed world navigation. After the creation, the resultant values will be their names (there may be more than one destination node created) with termination state thru, and the next steps will be associated with the nodes reached, starting in them. If the operation fails, its resultant value will be nil and control state fail in the node it started. After the node(s) successful removal operation, the resultant value in the starting node will be the same as before and control state thru.

create—starting in the current world position, creates either new virtual link-node pairs or new isolated nodes. For the first case, the rule is supplied with names and orientations of new links and names of new nodes these links should lead to, which may be multiple. For the second case, the rule has to use modifier direct indicating direct nodes creation, i.e. without links to them. If to use modifiers existing or passed for the link-node creation hinting that such nodes already exist (also if nodes are given by addresses, always indicating their existence) only links will be created to them by create.

linkup—just simplifies the latest rule, creating only links with proper names from the current node to the already existing nodes, without the need to use modifiers existing or passed. However, still using modifier passed may help us narrow direct search of the already existing nodes.

delete—removes links together with nodes they should lead to, starting from the current node. Links and nodes to be removed should be either explicitly named or represented by modifiers any or all. Using modifier direct instead of link name together with node name will allow us to remove such node (or nodes) from the current node directly. In all cases, when a node is deleted, its all links with other nodes will be removed too.

unlink—removes only links leading to neighboring nodes where, similar to the previous case, they should be explicitly named or modifiers any or all used instead. The resultant values on the rule will be represented by these node names, with states thru in them, similar to hop and linkup operations. The next scenario step will start in these neighboring nodes.

The above mentioned creation rules, depending on the implementation, can also be used in a broader sense and scale, as *contexts* embracing arbitrary scenarios and influencing hop operations within their scope (the same scenarios will be capable of operating in both creation and deletion mode with them).

4.5.3   Echoing

The rules of this class use terminal control states and terminal values from the embraced scenario (which may be remote) to obtain the resultant state and value in the world point it started, also being it's *terminal* point (from which the rest of the scenario, if any, will develop). The usual resultant control state for these rules is thru (fail occurs only when certain terminal values happen to be unavailable or result unachievable, say, as division by zero). Depending on the rule's semantics, the resultant value can be compound, like a list of values, which may be nested.

state—returns the resultant generalized state of the embraced SGL scenario upon its completion, whatever its complexity and space coverage. This state being the result of the ascending fringe-to-root generalization of terminal states of the scenario embraced, where states with higher power (from max to min, as already mentioned in the previous chapter: fatal, thru, done, fail) dominate in this potentially distributed and parallel process. The resultant state returned is treated as the *resultant value* on the rule, the latter always terminating with own control state thru, even in the case of resultant fatal, thus restricting the latter spreading by echo rules. (Another restriction of influence of fatal by a special rule will be explained later.)

order—returns an ordered list of final values of the scenario embraced corresponding to the order of launching related branches rather than the order of their completion. For parallel branches these orders may, for example, relate to how they were activated, possibly, with the use of time stamping upon invocation.

rake—returns a list of final values of the scenario embraced in an arbitrary order. This order may, for example, depend on the order of completion of branches; it can also be influenced by peculiarities of the echoing collection procedure of the results.

sum—returns the sum of all final values of the scenario embraced.

count—returns the number of all resultant values associated with the scenario embraced, rather than values themselves as by the previous rules.

first, last, min, max, random, average—return, correspondingly, the first, the last, minimum, maximum, randomly chosen, or average value from all terminal values returned by the scenario embraced, where first and last will depend on ordering of the results with details similar to the rule order above.

quickest—results in first in time returned value from the terminal ones of the scenario embraced.

element—returns the value of an element of the list on its left operand by index or content (see corresponding usage rules later) given by the right operand. If the right operand is itself a list of indices/contents, the result will be a list of corresponding values from the left operand. If element is used within the left operand of assignment, instead of returning values it will be providing an access to them, to be updated.

sortup, sortdown return an ordered list of values produced by the operand embraced, starting from maximum or minimum value and terminating, correspondingly, with minimum or maximum one.

reverse—changes to the opposite the order of values from the embraced operand.

add, subtract, multiply, divide, degree—perform the corresponding operations on two or more operands of the scenario embraced. If the operands represent multiple values as lists, these operations are performed between the peer elements, with the resulting value being multiple too.

separate—separates the left operand string value by the string at the right operand used as a delimiter in a repeated manner for the left string, with the result being the list of separated values. If the right operand is a list of delimiters, its elements will be used sequentially and cyclically unless the string at the left is fully partitioned. If the left operand represents a list of strings, each one is separated by the right operand as above, with the resultant lists of separated values merged into a common list in the order they were received.

unite—integrates the list of values at the left (as strings or to be converted into strings automatically) by a repeated delimiter as a string (or a cyclic list of them) at the right into a united string.

attach—makes the resultant string by connecting the right string operand to the end of the left one. If operands are lists with more than one element, the attachment is made between their peer elements, receiving the resultant list of united strings. This rule can also operate with more than two operands.

append—forms the resultant list from left and right operands, appending the latter to the end of the former, where both operands may be lists themselves. More than two operands can be used too.

common—returns intersection of two or more lists as operands, with the result including same elements of all lists, if any, otherwise nil.

withdraw—its result will be the first element of the list provided by the embraced operand, with this element also simultaneously withdrawn from the list. This rule can work with more than one element by adding another operand providing the number of elements to be withdrawn and represented as the result.

access—returns an internal access (which can be recorded, say, in a variable) to all terminal positions of the embraced scenario, which can be used to reenter them most efficiently afterwards (on internal system level). This reentry may be performed by the rule follow described before.

invert—changes the sign of a value or orientation of link to the opposite, producing no effect on zero or non-oriented operands.

4.5.4 Verification

These rules provide control state `thru` or `fail` reflecting the result of certain verification procedures, also `nil` as own resultant value, while remaining in the same world positions after completion.

equal, nonequal, less, lessorequal, more, moreorequal, big-ger, smaller, heavier, lighter, longer, shorter—make comparison between left and right operands, which can represent information or physical matter, or both. In case of vector operands, state `thru` appears only if all peer values satisfy the condition set up by the rule (except `nonequal`, for which even a single non-correspondence between peers will result in `thru`). The list of such rules can be easily extended for more specific applications, if supported properly on implementation level.

empty, nonempty—checks for emptiness (i.e. non-existence, same as `nil`) or non-emptiness (existence) of the resultant value obtained from the embraced scenario.

belongs, notbelongs—verifies whether the left operand value (single or a list) belongs as a whole to the right operand, potentially a list too.

intersects, nonintersects—verifies whether there are common elements (values) between left and right operands, being generally lists. More than two operands can be used for this rule too, with at least a same single element to be present in all of them to result in `thru`.

4.5.5 Assignment

This class of rules assigns the result of the right scenario operand (which may be arbitrarily remote, also as a list of values) to the variable or set of variables directly named or reached by the left scenario operand, which may be remote too. The left operand can also provide pointers to certain elements of the reached variables which should be changed by the assignment rather than the whole variables (see the rule `element` mentioned above). These rules will leave control in the same world position they've started, its resultant state `thru` if assignment was successful otherwise `fail`, and the same value as assigned to the left operand. There are two options of the assignment.

assign—assigns the same value of the right operand (which may be a list) to all variables accessed (or their elements pointed) by the left operand. If the right operand is represented by `nil` or empty, the left operand variables as a whole (or only their elements pointed) will be removed.

assignpeers—assigns values of different elements of the list on the right operand to different variables (or their pointed elements) associated with the destinations reached on the left operand, in a peer-to-peer mode.

4.5.6 Advancement

Rules of this class organize forward or "in depth" advancement in space and time. They can work in synchronous or asynchronous mode using modifiers synchronous or asynchronous (the second one optional as asynchronous is the default mode).

advance—organizes stepwise advancement in physical, virtual, executive or combined spaces, also in a pure computational space while staying in the same world nodes (thus moving in time only). For this, the embraced SGL scenarios-operands are used in a sequence, as written, where each new scenario applies from all terminal world points reached by the previous scenario (these points may happen to be the same as before if only computations took place). The resultant world positions and values on the rule are associated with the final steps of the final scenarios on the rule. And the rule's resultant state is a generalization of control states associated with its final steps. The frontal variables, if any, are being inherited at new steps from the preceding steps (with their copies removed from the previous positions), thus moving from one step to another, and between the scenario operands, being also replicated if multiple steps emerge from a previous step. If no final step occurs with states thru or done, the whole advancement on this rule is considered as failed (with generalized state fail), resulting in no possibility to continue the scenario evolution in this direction. On default or with modifier asynchronous, the sequence of scenarios develops in space and time independently in different directions, and different operand scenarios in the sequence may happen to be active at the same time. With the use of synchronous modifier, all invocations of every new scenario in their sequence can start only after full completion of all invocations of the previous scenario.

slide—works similar to the previous rule unless the next scenario fails to produce resultant state thru or done from some world node; in this case the next scenario from their sequence will be applied from the same starting position, and so on. The resultant world nodes and values in them will be from the last successfully applied scenario (not necessarily the same in their sequence when independently developing in different directions). The results on the whole rule, in their extreme, may even happen to correspond to the existing results in the node the rule started (including node's position) before the rule's application, with state thru always being the resultant state in any cases. Both synchronous and asynchronous modes of parallel interpretation of this rule, similar to the previous rule advance, can be possible, where in the synchronous case different scenarios can start only after full completion of the previous ones.

repeat—invokes the embraced scenario as many times as possible, with each new iteration taking place from all final positions with state thru reached by the previous invocation. If no final steps of the scenario invocation completed with state thru, the starting position from which this latest iteration failed together with its value will be included into the set of final positions and values on the whole rule (and this set may have positions from different iterations). Similar to the previous rule `slide`, in the extreme case the final set of positions on the whole rule may happen to contain only the position from which the rule started, with state `thru` and value it had at the beginning. By supplying additional numeric modifier to this rule, it is possible to explicitly limit the number of allowed repetitions of the embraced scenario (of course, the operand scenario may be organized to properly control the needed number of iterations itself, but with additional modifier this may be more convenient is come cases). Both synchronous and asynchronous modes of parallel interpretation of this rule, similar to the previous rules `advance` and `slide` are possible. In the synchronous mode, at any moment of time only the same scenario iteration can develop in a potentially distributed space-time continuum, whereas in the asynchronous case these may happen to be different iterations working in parallel.

`align`—causes coordination (alignment, synchronization) with other such rules which may be in other scenario branches; all will allow continuation of the subsequent parts of the scenario only after the coordination takes place. This rule may have identification and usage parameters which will be exemplified in other chapters.

4.5.7 Branching

These rules allow the embraced set of scenario operands to develop "in breadth", each from the same starting position, with the resultant set of positions and order of their appearance depending on the logic of a concrete branching rule. Branching may be static and explicit if we have a clear set of individual operand scenarios separated by comma. It can also be implicit and dynamic, as explained later. For all branching rules that follow, the frontal variables associated with the starting position will be replicated together with their contents, with the copies obtained developing independently within different branches. The original variable will be removed from the starting position then. Details of this replication if the variable holds physical matter rather than information can depend on the application and implementation peculiarities.

`branch`—most general variant with logical independence of scenario operands from each other, and any possible order of their invocation and development from the starting position (from strictly sequential to fully parallel, and from chaotic to absolutely ordered). The resultant set of positions and associated values will unite all terminal positions and values on all scenario operands involved, and the resultant control state on the whole rule being the generalization of generalized states on all scenario branches.

`sequential`—organizing strictly sequential invocation of all scenario operands, regardless of their resultant generalized control states, and launching the next scenario only after full completion of the previous one. The resultant set of positions, values, and rule's control state will be similar to `branch`.

parallel—organizing fully parallel development of all scenario operands from the same starting position (at least as much as this can be achieved within existing environment, resources, and implementation). The resultant set of positions, values, and rule's control state will be similar to the previous two rules.

if—usually has three scenario operands. If the *first* one results with generalized termination state thru or done, the *second* scenario is activated otherwise the *third* one will be launched. The resultant set of positions & associated values will be exactly the same as achieved by the second or third scenarios after their completion. If the third scenario is absent and the first one results with fail, the resultant position will be the one the rule started from, with state thru and value it had at the start. If only a *single* operand (i.e. the first one) is under the rule, it will also result with its starting position, initial value in it, and state thru, regardless of the generalized termination state of this single operand, its positions reached and values in them (all these will be ignored for the further scenario development, if any).

or—allows *only one* operand scenario in their sequence (not specifying which, may be any) with the resulting state thru or done to be registered as successful and resultant, with the resulting positions & associated values on it to be the resulting ones on the whole rule. The activities of all other scenario operands and all results produced by them will be cancelled. If no branch results with thru or done, the rule will terminate with fail and nil value. Used in combination with the previous rules sequential and parallel, it may have the following more clarified and detailed options.

orsequential—launches the scenario operands in a strictly sequential manner, one after the other as they are written, waiting for their full completion before launching the next one, unless the first one replying with generalized state thru or done, providing the result on the rule as a whole. Invocation of the remaining scenarios in the sequence will be skipped.

orparallel—activates all scenario operands in parallel from the same current position, with the first one in time replying with generalized thru or done being registered as the resultant branch for the rule. All other branches will be forcefully terminated without waiting for their completion (or just ignored, depending on implementation, which in general may not be the same due to side effects when working with common resources).

The resultant scenario in all three cases above provides its final set of positions with values and states in them as the result on the whole rule. If no scenario operand returns states thru or done, the whole rule will result with state fail in its starting position and nil as resultant value.

and—activates each scenario operand from the same position, demanding all of them to return generalized states thru or done. If at least a single operand returns generalized fail, the whole rule results with state fail and nil value in the starting position while forcefully terminating the development of all other branches, which may still be in progress. If all operand scenarios succeed, the resulting set of positions unites all resultant positions on all operands with their associated values. Combining the rule with rules sequential and parallel (as we did for or) clarifies their activation and termination order, as follows. (These two options can, in principle, produce differing general results if different scenario operands work in intersecting domains and share intermediate results.)

andsequential—activates each scenario operand from the same position in the written order, terminating the rule when first one results with fail, while ignoring other operands and removing all results produced by this and all previous operands.

andparallel—activates each scenario operand from the same position, terminating the rule when the first one in time results with fail, while aborting all other operands activity and removing all results produced by the current one.

choose—chooses a scenario branch in their sequence *before* its execution, using certain parameters among which, for example, may be its numerical order in the sequence (or a list of such orders to select more than one branch). This rule can also be aggregated with other rules like first, last, random, or any clarifying the branch to be chosen. The resultant set of positions, their values and states will be taken from the branch(es) chosen.

firstrespond—selects the first branch in time replying its complete termination, regardless of its generalized termination state, which may happen to be fail too, even though the other branches (to be forcefully terminated afterwards) could respond later with thru or done. The set of positions on this selected branch and their associated values (if any) will be taken as those for the whole rule. This rule assumes that different branches are launched independently and in parallel. But it differs fundamentally from the rule orparallel as the latter selects the first in time branch replying with success (i.e. thru or done) for which, in the worst case, all branches may need to be executed in full to find the branch needed. A modification of this rule my have an additional parameter establishing, for example time limit within which replies are expected or allowed from branches (where there may be more than one branch as the result), otherwise failure if no branch responded in time.

cycle—repeatedly invokes the embraced scenario from the same starting position until its resultant generalized state remains thru or done, where on different invocations same or different sets of resultant positions with different values may emerge. The resultant set of positions on the rule will be an integration of all positions on all successful scenario invocations with their values. If no invocation of the embraced scenario succeeds, the resultant state fail in the starting position and nil value will emerge.

loop—differs from the previous rule in that the resultant set of positions on it being only the set produced by the *last* successful invocation of the embraced scenario (the rule will terminate, as before, with fail and nil in the starting position if no invocation succeeds).

sling—invokes repeatedly the embraced scenario until it provides state thru or done, resulting in the same starting position with state thru and its associated value when the last iteration results with fail.

whirl—endlessly repeating the embraced scenario from the starting position regardless of its success or failure with no resultant positions or values produced.

External forceful termination of this construct may be needed, like using first in time termination of a competitive branch (say, under higher-level rule `orparallel`).

It could also be possible to set a limit on the number of repetitions (or duration time) in these cycling-looping-slinging-whirling rules—by supplying them with an additional parameter restricting the repeated scenario invocations.

`split`—performs, if needed, additional static or dynamic partitioning of the embraced scenario to different branches, especially in complex and not clear at first sight cases, all starting from the same current position. It may be used alone or in combination with the above mentioned branching rules, preparing separate branches for the latter. Some examples follow.

- If `split` embraces explicit branches separated by commas, it does nothing as the branches are already declared.
- It the embraced single operand represents broadcasting move or hop (creative or destructive including) in multiple directions, the branches are formed from all possible variants of elementary moves or hops, *before* their execution.
- If the rule's operand is an arbitrary scenario (not belonging to the two cases above), the branches are formed *after* its completion, where each position reached (with associated values) starts a new branch.
- If an arbitrary scenario terminates with a single or multiple positions which have multiple values associated with them (i.e. lists), each constituent value in these lists starts an individual branch, becoming its sole value.

`fringe`—being the most general variant of splitting for any scenario after its execution, is considering all final positions reached by the scenario as individual branches. It may also have additional parameters helping us to select or reject the received branches as candidates for a further scenario evolution (possibly, with involvement of both forward and echo operations over the control hierarchy produced by the scenario, for making proper decisions). Examples explaining the use of this rule can be found in the following chapters.

4.5.8   Transference

This class of rules organizes different control or data transference.

`run`—transfers control to the SGL code treated as a procedure and represented as a result of invocation of the embraced scenario (which can be of arbitrary complexity and space coverage). The procedure (or a list of them) obtained and activated in such a way will produce a set of world positions with associated values and control states as the result on the rule, similar to other rules.

`call`—transfers control to a code produced by the embraced scenario which may represent activation of external systems (including those working in other formalisms), with resultant world position being the same where the rule started, value in it corresponding to what has been returned from the external call, and state `thru` if the call was successful, otherwise `fail`.

input—provides input of external information or physical matter (objects) on the initiative of SGL scenario, resulting in the same position but with value received from the outside. The rule may have an additional argument clarifying a particular external source from which the input should take place. The rule extends possibilities provided by reading from environmental variable `IN` explained before.

output—outputs the resultant value obtained by the embraced scenario, which can be multiple, with the same resultant position as before and associated value sent outside (for virtual data only). The rule may have an additional pointer to a particular external sink. The rule extends possibilities provided by assignment to the previously explained environmental variable `OUT`.

transmit—represents a variant of `output` for specific applications, say, involving long distance radio communications and broadcasting features, with potentially multiple addresses. It may have additional parameters clarifying the action needed.

send—staying in the current position associated with physical, virtual, executive (or combined) node, transfers information or matter obtained by the scenario on the first operand to other similar node given by name, address or coordinates provided by the second operand, assuming that a companion rule `receive` is engaged there. The rule may have an additional parameter setting acceptable time delay for a consumption of this data at the receiving end. If the transaction is successful, the resultant position will be the same where the rule started with state `thru` and value sent (virtual only) otherwise `nil` and state `fail`.

receive—a companion to rule `send`, naming the source of data to be received from (defined similarly to the destination node in `send`). Additional timing (as a second operand) may be set up too after expiration of which the rule will be considered as failed. In case of successful receipt of data, the rule will result in the same world position with the value obtained from `send` and state `thru`, otherwise with `nil` and state `fail`.

4.5.9   Timing

These rules are dealing with conditions related to time for the scenarios they embrace, with currently only two being used, as follows.

sleep—establishes time delay defined by the embraced scenario operand, with no activities in the meantime by this particular scenario branch. The starting position and its existing value will be the result on the rule after the time passed, with state `thru`. Such time delay of the related branch can also be achieved by assigning the current absolute time (received from environmental variable `TIME`) incremented by the delay value returned from the scenario embraced by `sleep`, to environmental variable `WHEN` described before.

allowed—sets time limit by the first operand for an activity of the scenario on the second operand. If the scenario terminates before time limit expires, its resultant positions with values and states will define the result on this rule. Otherwise the scenario will be forcefully aborted with state `fail` in the starting position as the rule's result.

4.5.10 Granting

These rules are granting certain capabilities and qualities, or setting constraints or restrictions to the scenarios they embrace, as follows.

contain—restricts the spread of destructive consequences caused by control state `fatal` within the ruled scenario. This state may appear automatically or can be assigned explicitly to environmental variable `STATE`, triggering emergent completion of all scenario processes and removal of data associated with the scenario. The resultant position will the one the rule started from, its value `nil`, and state `fail`. Without occurrence of `fatal`, the resultant positions, their values and states on the rule will be exactly the same as of the scenario embraced.

release—allows the embraced scenario develop free from the main scenario, abandoning bilateral control links with it, starting from the current position (the main scenario after the rule's activation "will not see" this construct any more). The released, now independent, scenario will develop using standard subordination and command and control mechanisms, as usual. For the main scenario, this rule will immediately result in its starting position with state `thru` and original value there.

free—differs from the previous case in that despite its independence and control freedom from the main scenario, as before, the embraced scenario is nevertheless obliged to return data obtained in its terminal positions if such a request has been issued by rules at higher levels.

blind—blocks the embraced scenario from engagement in further development after its completion, but retains the scenario's capability to reply to higher levels with values associated with final positions reached. This being equivalent to setting control state `done` in each terminal position of the embraced scenario.

stop—similar to the previous construct, but unlike `blind` it causes absolute termination in the current point, which cannot be revoked from higher levels (see sate `done` on which `blind` is based).

lift—removes blocking of further development caused by states `done` in terminal positions of the embraced scenarios (including the effect caused by previous rule `blind`), substituting them with `thru`, thus allowing further development from these positions by a subsequent scenario.

none—sets `nil` (or empty) as a returned value of the whole scenario embraced, with the rule resulting in the same starting position with sate `thru`.

stay—whatever the scenario embraced and its evolution in space, the resultant position will always be the same this rule started from, with the latest value in it and state thru. As can be seen, this rule differs from the previous one none only by its resultant value.

seize—establishes, or "seizes", an absolute control over the resources associated with the current virtual, physical, executive or combined node, blocking these from any other accesses and allowing only the embraced scenario to work with them, thus preventing possible competition for the node's assets which may lead to unexpected results. This resource blockage is automatically lifted after the embraced scenario terminates. The resultant set of positions on the rule with their values and states will be the ones from the scenario embraced (the latter may potentially be of any complexity and space-time coverage). If the node has already been blocked by another scenario exercising its own rule seize, the current scenario will be waiting for the release of the node. If more than two scenarios are competing for the node's resources, they will be organized in a FIFO manner at the node.

4.5.11 Type

These rules explicitly assign types to different constructs generally represented as strings (given explicitly or being the result of an arbitrary operand scenario with single or multiple elements). These rules result in the same positions the rule started, nil value and state thru (fail appears only if the string element does not satisfy certain constrains mentioned below).

global, heritable, frontal, nodal, environmental—allow different types of variables to have any identifiers (letter and/or digits only) rather than those restricted for self-identification, as explained before. These new names will continue represent the variables with their types in the subsequent scenario development to its full depth unless redefined by these rules. As regards environmental variables, their names differing from the standard ones and new kinds of such variables may need special adjustment with the implementation layer which is directly accessing corresponding physical or virtual resources.

matter, number, string, scenario—allow arbitrary strings (with letters, digits and some other characters but not violating the SGL syntax) obtained by the scenario embraced to represent corresponding values rather than using self-identifiable representations mentioned before (with automatic internal types conversion, if needed).

4.5.12 Usage

This class of rules explains how to use the information units they embrace.

address, coordinate, content, index, time, speed, name, place, center, range, doer, human, soldier, robot, nodes, links—explicitly clarify the purpose or usage of different values in other rules, adding flexibility to composition of SGL scenarios for which strict order of operands and presence all of them may be optional. The rules result in the same positions they've started with the values clarified by them.

unit—identifies the set of values produced by the embraced scenario as an integral unit (i.e. a list) for further processing. This may also be useful for hierarchical structuring of data, where elements within declared units may be other units themselves, and so on. The rule results in the same position it started with the value being the unit formed.

4.5.13 Application

Additional, custom, rules can allow SGL to be extended unlimitedly while embracing and embedding specifics of different application areas. They can be used similarly to other language rules while obeying established internal interpretation principles and unified command and control. These rules will, however, need extension of and adjustment to the underlying language interpretation system.

4.5.14 Grasp

This brings another level of recursion into the language structure where names of rules can themselves be determined by arbitrary scenarios, or *grasps* again, which can also allow rules (named explicitly or as results of scenarios) to be aggregated with each other into compound rules operating jointly over the same scenarios. Such an aggregation can also simplify the structure of scenarios by reducing enclosed into each other parentheses in them.

4.6 Conclusion

We have presented the full description of Spatial Grasp Language (SGL) including its recursive syntax and semantics of main constructs, which is sufficient for the language understanding and programming in it of various problems in large distributed systems. This detailed language description further clarifies and explains peculiarities of the Spatial Grasp Model described in the previous chapter, which is based on parallel navigation and grasping of distributed environments while bringing operations, control, and intermediate results directly into their points, also making exchanges between different points with these and other substances. SGL allows us to dynamically create spatial hierarchical control infrastructures in distributed spaces and make proper decisions at different conceptual and organizational levels without any dedicated central resources. SGL also directly operates with any remote data resulting from the spatial system coverage.

References

1. P. Sapaty, A brief introduction to the spatial grasp language (SGL). J. Comput. Sci. Syst. Biol. **9**(2) (2016)
2. P.S. Sapaty, Spatial grasp language for distributed management and control, in *MMC*, No. 3 (2016)
3. P. Sapaty, Spatial grasp language (SGL). Adv. Image Video Process. **4**(1) (2016)
4. P.S. Sapaty, Spatial grasp language (SGL) for distributed management and control. J. Robot. Networking Artif. Life **4**(2) (2016)
5. P.S. Sapaty, On possibilities of the organization of a direct intercomputer dialogue in ANALYTIC and FORTRAN languages, Publ. No. 74-29 (Inst. of Cybernetics Press, Kiev, 1974, in Russian)
6. P.S. Sapaty, A wave language for parallel processing of semantic networks. Comput. Artif. Intell. **5**(4) (1986)
7. P.S. Sapaty, The WAVE-1: a new ideology and language of distributed processing on graphs and networks. Comput. Artif. Intell. No. 5 (1987)
8. P.S. Sapaty, A brief introduction to the WAVE language, Report No. 3/93, Faculty of Informatics, University of Karlsruhe (1993)
9. P.S. Sapaty, Mobile Programming in WAVE. Math. Mach. Syst., No. 1, Jan–Mar 1998, Kiev, pp. 3–31, ISSN:1028-9763
10. P. Sapaty, *Mobile Processing in Distributed and Open Environments* (Wiley, New York, 1999)
11. P.S. Sapaty, M. Sugisaka, A language for programming distributed multi-robot systems, in *Proceedings of the Seventh International Symposium on Artificial Life and Robotics (AROB 7th ' 02)*, 16–18 Jan 2002, B-Com Plaza, Beppu, Oita, Japan, pp. 586–589
12. P. Sapaty, *Ruling Distributed Dynamic Worlds* (Wiley, New York, 2005)
13. P. Sapaty, *Distributed Technology for Global Control*. Lecture Notes in Electrical Engineering, vol. 37, Part 1, 3–24 (2009). doi:10.1007/978-3-642-00271-7_1
14. P. Sapaty, Meeting the world challenges with advanced system organizations, in *Informatics in Control Automation and Robotics*. Lecture Notes in Electrical Engineering, vol. 85, 1st edn. (Springer, New York, 2011)

Chapter 5
Main Spatial Mechanisms in SGL

5.1 Introduction

As follows from the definitions in Chaps. 3 and 4, any SGL operation is represented within a unified concept called *grasp*. We will discuss in more detail of how the grasp looks like when observed from outside rather than inside, and how different grasps can be integrated within SGL scenarios of any complexity and space and time coverage. We will be using improvised 3D graphical support in explaining and analysing different constructs which can shed more light on the language potential capabilities in expressing and solving problems in distributed environments.

The chapter describes the main spatial mechanisms of SGL with the help of so called "progress points", or *props*, as integration of emerging control with the locations reached by it in navigated spaces. Different mechanisms activated by SGL rules for both breadth and depth evolution of spatial scenarios are exhibited and explained including parallel, sequential, synchronous, asynchronous, alternative, cyclic, looping, whirling, sliding, and other variants. All these, especially if used together, are allowing for an overall vision, investigation, analysis, and control of large distributed spaces with clear and compact linguistic expressions.

Some of these mechanisms have been already discussed in [1–11], and the current chapter presents their extended repertoire with more detailed explanations.

5.2 Progress Points or Props

As described in the previous chapters, SGL scenarios are developing in *steps* which relate to the world positions they have started from, also reached as destinations. These positions always associate with the scenario control which is staying in them temporarily and moving in between. For a further clarification of this space and control association, we are introducing here a notion of *progress point*, or *prop*

© Springer International Publishing AG 2017
P.S. Sapaty, *Managing Distributed Dynamic Systems with Spatial Grasp Technology*, Studies in Computational Intelligence 690,
DOI 10.1007/978-3-319-50461-2_5

(the latter as an abbreviation only, differing from traditional meaning of the word "prop"). Our prop will formalize and individualize these scenario steps, as the latter, if multiple and parallel, may happen to be associated in time with the same world positions, thus losing spatial identity. And props, as integration of the unique scenario control points with current world positions (same or different), will always be unique too.

5.3   Single Grasp Representation

Main developmental features of a *single grasp* (let us call it G) of any complexity (from just declaring a constant or variable to any scenario covering arbitrary operations in/on arbitrary territory) are shown in Fig. 5.1.

From the Starting Prop (SP) associated with a current point in physical, virtual, executive or combined world, a grasp evolves and covers certain world region referred to as World Coverage (or WCO)—performing operations, obtaining data, making decisions, also changing this world if needed.

The final result of this spatial evolution is considered to be in the area represented and covered by the Resultant Set of Props (RSP), each linked with the same or different world points reached. After completion of the grasp, its Resultant Control State (RCS) obtained in a parallel echo process merging termination states from final props belonging to RSP is available, if requested, in the SP, and may be taken into account for making decisions at higher levels controlled by the outer rules.

Also, upon a request from higher levels, the RSP's Associated Values (AV) can be lifted, collected, merged, processed, etc., generally hierarchically and in parallel within the spatial echo process, and returned to SP for further consideration as a Resultant Value (RV) on the whole grasp. Final states (FS) associated with props in FSP can only be thru or done, confirming the very existence of these props.

There may be different relations between SP, WCO, and RSP as between sets, shown in Fig. 5.2 (RSP depicted as shaded), which may result from the semantics of different SGL rules described in Chap. 4.

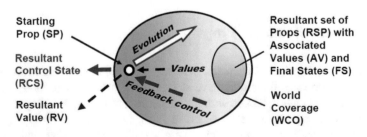

Fig. 5.1 Single grasp general features

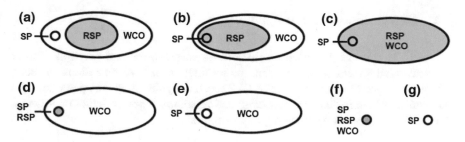

Fig. 5.2 Possible relations between SP, WCO, and RSP

Let us explain these variants.

(a) Any RSP possible, with any WCO, where WCO considered greater than RSP, and SP and RSP do not intersect (this case actually corresponds to the situation shown in Fig. 5.1).
(b) This case differs from the previous one in that RSP may include SP, the latter thus belonging to the final results too.
(c) WCO and RSP being the same, making the full space coverage as total final result (also including SP as always part of WCO).
(d) SP being the same as RSP, with any WCO, thus associating final result only with SP.
(e) No RSP achieved starting from SP with any WCO possible, thus no final results obtained and no further scenario development.
(f) SP, WCO, and RSP, all three, are the same, symbolizing the whole grasp operating in the starting point only, with final results in it too.
(g) Only SP, which means no space coverage, no final results achieved, and no further development possible.

In cases where no RSP achieved, like in Fig. 5.2e, it does not mean, however, that no intermediate results have been produced, left and recorded throughout WCO during the navigation and coverage process (including world creation and modification), as the grasp can represent arbitrary complex spatial algorithm. And these results can be further accessed by other scenarios or other branches of the current one, for which they may be used in intermediate or final results.

The mentioned grasp's general features (i.e. SP, WCO, RSP, AV, FS, RCS, and RV) can represent any SGL scenarios of any complexity within the same framework shown in Fig. 5.1. This allows us to effectively combine and integrate different grasps within arbitrary complex spatial scenarios reflecting any parallel and distributed algorithms over any physical and virtual spaces using simple and standard rules, as will be shown in the rest of this chapter.

5.4   Depth Mode Space Navigation

We will fist consider composition that allows us move forward in virtual and/or physical environment or just in time performing successive operations in them, where next operations are starting in RSP reached by the previous operations. We will consider here *advancement*, *sliding* and *repetition* types of depth, or forward, composition of grasps representing any operations.

5.4.1   Advancement

The most common depth mode of composition and development of grasps can be represented by a sequence of arbitrary grasps under the rule advance with its different modifications.

5.4.1.1   General Advancement

General advancement operating with grasps (let them be three) G1, G2, G3 can be expressed with the use of rule advance as follows:

$$advance(G1, G2, G3).$$

Without details of implementation it is shown in Fig. 5.3 where each subsequent grasp is applied from all props of RSP of the previous grasp (i.e. those having state thru, not done). The RSP (and AV) on the whole rule will be determined by the RSP of last applied grasp (i.e. G3), more correctly, by all its invocations which could start from more than one prop produced by the previous grasp (i.e. G2, by all its invocations too, and so on, back to the single SP). In Fig. 5.3, G2 and G3 are therefore symbolizing all invocations of these two grasps rather than their single copies (but not G1 reflecting a single copy starting from SP).

The advancement mode of navigation can be organized asynchronously or in a synchronized manner by using explicit modifiers asynchronous or synchronous, as explained below. With these omitted, on default, the rule advance operates in asynchronous mode. More details on these two options follow.

5.4.1.2   Asynchronous Advancement

This can be written as one of the following options, due to SGL syntax flexibility:

* advance(asynchronous)(G1, G2, G3)
* (advance, asynchronous)(G1, G2, G3)
* advance(asynchronous, G1, G2, G3)

Fig. 5.3 Advancement
composition of grasps

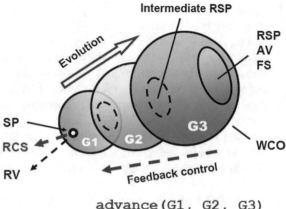

advance(G1, G2, G3)

- advance_asynchronous(G1, G2, G3)
- advance_async(G1, G2, G3)—shortened
- advance(G1, G2, G3) on default.

A detailed development of this mode of advancement is shown in Fig. 5.4.

Within this mode, each new grasp is developing from every prop (with state thru) from RSPs of the previous grasp invocations, upon their earliest availability, without waiting for completion of other invocations. Grasps in their sequence, except the last one, do not report their termination states to the previous grasps (also back to the SP). The resultant state on the rule and resultant values will be determined exclusively by the RSPs of the last grasp (i.e. G3) invocations, and can be echoed to the SP if requested. If no RSP on the last grasp reached, the rule will terminate with resulting state fail and nil value in the SP.

The asynchronous mode effectively allows for moving of SGL code in distributed space, with used code parts automatically removed after each stage of development and the rest pushed further, as follows (also shown in Fig. 5.4):

Fig. 5.4 Asynchronous
advancement

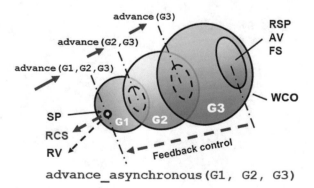

advance_asynchronous(G1, G2, G3)

1. Applied in SP: `advance(G1, G2, G3)`, executing `G1` while carrying the remaining code `advance(G2, G3)` accompanying the spatial evolution of G1 and replicating in case of branching in G1 until it reaches RSP1.
2. Applied in RSP1: `advance(G2, G3)`, executing `G2` while carrying the remaining code `advance(G3)` accompanying the evolution of G2 und replicating if G2 is branching until it reaches RSP2.
3. Applied in RSP2: `advance(G3)`, executing `G3` (with no remaining code to transfer).
4. Nothing to apply in RSP3, which keeps the final result, echoing AV and FS back to SP (if requested from a higher level only) with getting RV and RCS in SP using (possibly, self-optimized) paths created by the activity of `G1`, `G2`, and `G3`.
5. If no RSP3 obtained, the rule will result in SP with `fail`.

5.4.1.3 Synchronous Advancement

The synchronous advancement can be written as one of:

- `advance(synchronous) (G1, G2, G3)`
- `(advance, synchronous) (G1, G2, G3)`
- `advance(synchronous, G1, G2, G3)`
- `advance_synchronous(G1, G2, G3)`
- `advance_sync(G1, G2, G3)`—shortened.

Development of this mode of advancement is shown in Fig. 5.5.

Under synchronous advancing each new grasp in their sequence cannot start unless all invocations of the previous grasp report their full completion, altogether, with generalized state `thru` resulting in SP. This reporting needs control feedback (via echoed state generalization, back to SP) from every grasp termination, after all of which (and only) the next grasp in their sequence will be launched from RSPs of all invocations of the previous grasp.

Fig. 5.5 Synchronous advancement

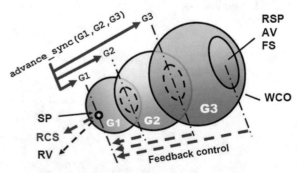

`advance_synchronous(G1, G2, G3)`

Fig. 5.6 Synchronous
advancing with independent
code movement

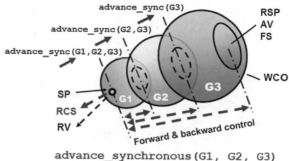

advance_synchronous(G1, G2, G3)

There may be two options of implementation of the synchronous advance.

Option 1. SP keeps full code of the rule with gasps embraced, sending to RSP of each grasp the next grasp code only after full completion of all grasp invocations (as shown in Fig. 5.5). The sequence of steps for this case, each originating in SP will be as follows:

1. Deploy advance_synchronous(G1, G2, G3) in SP.
2. Activate G1 in SP.
3. Wait for reply from RSP1 with generalized state thru, otherwise stop in SP with state fail.
4. Broadcast G2 code to RSP1 and activate it there (where broadcasting may use self-optimized paths to RSP1 created by the work of G1).
5. Wait for reply from RSP2 in all G2 invocations with getting generalized state thru in SP, otherwise stop in SP with fail.
6. Broadcast G3 code to RSP2 in all G2 invocations and activate it there.
7. Wait for reply from RSP3 in all G3 invocations while echoing their AV and FS back to SP getting RCS and RV of the whole rule.
8. If no RSP3 obtained, the rule will result in SP with fail.

Option 2. The code on the rule with embraced grasps is moving from grasp to grasp invocation, as in the previous asynchronous case while omitting the used grasps, but the start of execution of each new grasp from all RSPs of the previous grasp is allowed only in a centralized manner by permission from SP, as shown in Fig. 5.6. Instead of broadcasting the scenario code for each grasp invocation, in this case only invocation permissions will be broadcast, as the needed grasp code is already present in the starting props for the new grasp.

5.4.2 Sliding

The rule slide resembles rule advance and can be implemented with all variations discussed for advance, but allows us to use the next grasp in their sequence from any prop where invocation of the previous grasp resulted with fail, and so on,

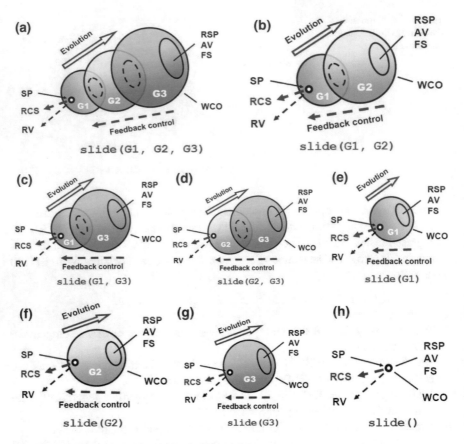

Fig. 5.7 Possible cases of the work of rule slide

until all grasps are tried. The RSP on the rule is an amalgamation of RSPs from all invocations of the last successful grasps (not necessarily the same) or equals to SP with state thru if no grasp succeeded from it. This rule can develop with modifiers asynchronous or synchronous, and on default is equivalent to the use of asynchronous. Figure 5.7a–h shows all possible cases of the evolution of slide (G1, G2, G3), which can all be potentially present under the same rule when developing independently in different directions within the space-time continuum.

5.4.3 Repetition

The repetition mode, represented by rule repeat, allows for potentially unlimited number of usage of the embraced grasp in navigation of the distributed space-time continuum.

Fig. 5.8 Repetitive
navigation

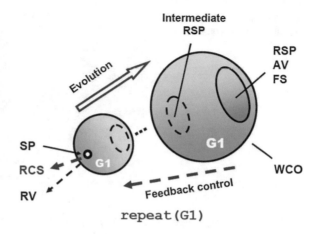

repeat(G1)

5.4.3.1 General Repetition

The most general representation of repetition is as follows, where G1 can stand for
an arbitrary SGL scenario, or grasp:

$$\text{repeat(G1)}$$

Under this rule, the embraced grasp develops potentially unlimited number of
times where each new invocation starts from RSPs of all invocations of the previous
grasp, as shown in Fig. 5.8.

The RSP with AV and FS on this rule can be fixed and obtained only when it
stops as a whole, and this will include the resultant props with state done or the
props from which the latest invocation of the grasp resulted with generalized fail
(which will be changed in them for thru to allow the rest of the scenario, if any, to
proceed from the related world positions). For the both cases and their combinations
(as repeat can develop differently in different space and time directions) not
necessarily the same iterations of the grasp will participate in the rule's final
RSP. In the extreme case, the rule's RSP may contain only its SP (with RCS as
thru) if no success on the grasp's invocation, even a single one, has been
achieved.

There may be two modes of this rule's operation: asynchronous and synchronous, similar to rules advance and slide discussed before.

5.4.3.2 Asynchronous Repetition

The asynchronous mode can be syntactically set up in a variety of ways, similar to
the previous rules, as follows:

- repeat(asynchronous) (G1)
- (repeat, asynchronous) (G1)

Fig. 5.9 Asynchronous
repetitive advancement

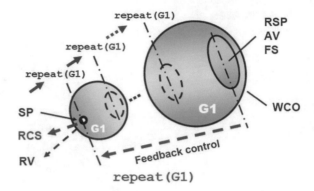

- `repeat(asynchronous, G1)`
- `repeat_asynchronous(G1)`
- `repeat_async(G1)`—shortened
- `repeat(G1)`—on default.

Development of asynchronous repetition is shown in Fig. 5.9.

After termination of each iteration, the whole `repeat` rule with embraced grasp is applied again fully independently from RSPs of the previous iterations with full rule's code freely moving and replicating between successive RSPs. The rule echoes as a whole with final RSP and AV and FS to the SP (if this is required by higher control levels) after all iterations exhaust their capabilities.

5.4.3.3 Synchronous Repetition

This synchronous mode, shown in Fig. 5.10, can also have different syntactic representations, as follows:

- `repeat(synchronous) (G1)`
- `(repeat, synchronous) (G1)`
- `repeat(synchronous, G1)`
- `repeat_synchronous(G1)`
- `repeat_sync(G1)`—shortened.

Developing from SP, the invocation of a new iteration waits for full completion of all copies of the previous iteration (this indication can be achieved by echoing termination states of each iteration back to the SP). Only after this, the grasp can be applied to RSPs of the previous iteration again (excluding the ones from which the grasp already failed, but which participate in the final RSP on the rule).

Similar to rule `advance`, the grasp code for each step (here iteration) can be broadcast from SP directly to RSP of the previous iteration (this case corresponding to Fig. 5.10). It can also propagate feely in space between RSPs of consecutive

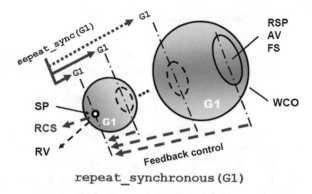

Fig. 5.10 Synchronous repetitive advancement

repeat_synchronous(G1)

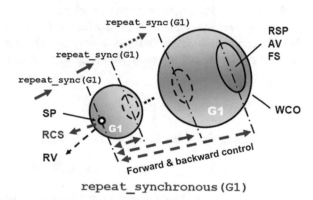

Fig. 5.11 Synchronous repetitive advancement with independent code moving

repeat_synchronous(G1)

iterations along the grasp executions without involving SP, as in Fig. 5.11. The SP, however, will have to retain its role as a command and control center coordinating strict stepwise development of the rule, exchanging only commands and confirmations but not SGL scenario code itself, which is already at the places reached.

5.5 Breadth Mode Navigation

Within the breadth composition of grasps each of them is applied from the same SP, with the order of their invocation and definition of final results determined by the semantics of concrete rules described in Chap. 4. We will shed here some more light on peculiarities of the breadth type composition rules, supporting the narrative with detailed graphical representations.

5.5.1   General Branching

The breadth branching composition of grasps in its most general, loose form can be written as follows (let there be three grasps only for simplicity, as before):

$$\mathrm{branch}(G1, G2, G3)$$

It is shown in Fig. 5.12 where all grasps in the group are evolving from the same starting prop.

With this general branching, all the constituent grasp-branches are developing independently, their order of invocation and termination can be arbitrary, and resultant props of all branches will contribute to the united RSP with data values and final states (i.e. thru or done) associated with them. The RCS on the rule will be a merge (i.e. the maximum) among final states on all resultant props. In case no branch replies with generalized thru or done, the result on the whole rule will be represented in the starting prop with state fail and nil value.

5.5.1.1   Sequential Branching

Sequential branching can be set up as follows:

$$\mathrm{sequence}(G1, G2, G3)$$

This rule, in comparison with the loose branching by branch, clarifies the exact order of invocation of constituent grasps, which should be strictly sequential, where a new branch can start from the same SP only after the previous branch has been fully completed, regardless of its result and own RCS (except when state fatal emerges), as shown in Fig. 5.13.

Fig. 5.12 General branching

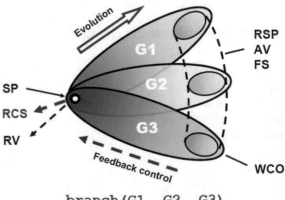

branch(G1, G2, G3)

Fig. 5.13 Sequential
execution of branches

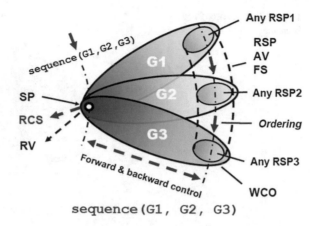

sequence (G1, G2, G3)

The RSP on the whole rule with its AV and FS will be uniting RSPs on all
braches with their AV and FS, similar to rule branch.

5.5.1.2   Parallel Branching

This can be set up by:

$$parallel(G1, G2, G3)$$

This rule, in comparison with loose branching by branch, activates all branches simultaneously and in parallel, as shown in Fig. 5.14, regardless of their
success or failure.

The RSP and AV on the rule will be uniting RSPs with their AVs and FS on all
braches, as for the rules branch or sequence. The final result on this rule may,

Fig. 5.14 Parallel execution
of branches

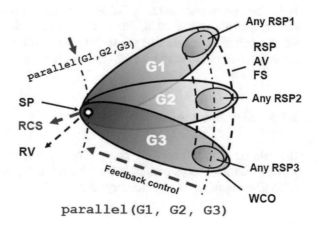

parallel (G1, G2, G3)

Fig. 5.15 IF-THEN-ELSE
branching

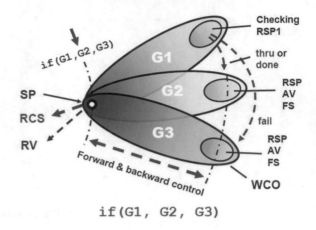

if(G1, G2, G3)

however, happen to differ from that obtained on branch or sequence due to
different order of operations which may be sharing common distributed resources.

5.5.2 IF-THEN-ELSE Branching

This kind of branching can be set up as:

$$if(G1, G2, G3)$$

The corresponding rule if, as already mentioned in Chap. 4, may have up to
three grasps as operands, as shown in Fig. 5.15, with G1, G2, and G3 written in a
sequence.

The rule operates as follows. If G1 results with generalized thru or done, G2
will be activated from the same starting prop, providing upon termination the RSP
on the whole rule with its associated values, otherwise G3 will be launched giving
its resultant props and values and states to the rule. If G3 is absent, the SP will be as
RSP on the rule in case G1 results with fail. And if both G2 and G3 are absent,
the rule will always result in the starting prop with state thru regardless of the
generalized termination state of G1 (except when state fatal happens within G1).

5.5.3 OR Branching

The rule or allows only one grasp with the resulting state thru or done to
provide resulting props and associated values and final states. With three grasps
again (their number can be arbitrary) it is written as:

Fig. 5.16 Sequential OR branching

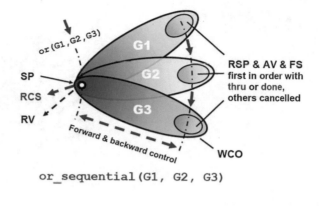

or_sequential(G1, G2, G3)

$$or(G1, G2, G3)$$

To clarify the grasps invocation order, the rule can be used in combination with the previous rules sequence and parallel.

5.5.3.1 Sequential OR Branching

This modification of the rule OR can be written in the following ways:

- (or, sequence) (G1, G2, G3)
- or_sequence(G1, G2, G3)
- or_seq(G1, G2, G3)—shortened
- or(G1, G2, G3)—by default.

The operation of sequential OR is depicted in Fig. 5.16.

The grasps are launched sequentially, waiting for their full completion before launching the next grasp, unless the first one in the sequence replies with generalized state thru or done providing the RSP for the rule with AV and FS while canceling invocation of the remaining grasps.

5.5.3.2 Parallel OR Branching

This OR rule option, shown in Fig. 5.17, can be written as follows:

- (or, parallel) (G1, G2, G3)
- or_parallel(G1, G2, G3)
- or_par(G1, G2, G3)—shortened.

All operand grasps are activated in parallel from the same current prop, with the first one in time with generalized state thru or done representing the RSP of the whole rule, while aborting all other branches which may still be active.

Fig. 5.17 Parallel OR
branching

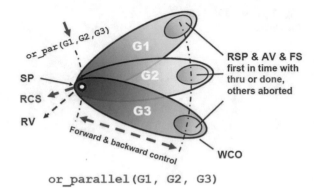

or_parallel(G1, G2, G3)

5.5.4   AND Branching

This type of branching represented by rule and demands all branches to terminate
with generalized state thru or done, with all contributing to the resultant set of
props with associated values and final states on the rule. If at least a single branch
replies with fail, the result of the whole rule will be fail too, also with nil value
in the starting prop. Both sequential and parallel options of this rule are available.

5.5.4.1   Sequential AND Branching

The sequential AND option can be expressed as follows:

- (and, sequence)(G1, G2, G3)
- and_sequence(G1, G2, G3)
- and_seq(G1, G2, G3)—shortened
- and(G1, G2, G3)—on default.

Its operation is depicted in Fig. 5.18.

Fig. 5.18 Sequential AND
branching

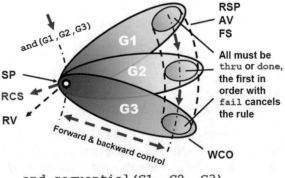

and_sequential(G1, G2, G3)

Fig. 5.19 Parallel AND
branching

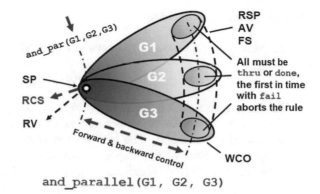

and_parallel(G1, G2, G3)

For this variant, the grasps are activated strictly sequentially, one after full completion of the other, and the first grasp in the sequence terminating with generalized `fail` cancels the whole construct with its resulting state `fail` too. Otherwise all branches contribute to RSP with their AV and FS to the whole rule.

5.5.4.2   Parallel AND Branching

This modification of the `and` rule can be written as follows, with its operation depicted in Fig. 5.19:

- `(and, parallel)(G1, G2, G3)`
- `and_parallel(G1, G2, G3)`
- `and_par(G1, G2, G3)`—shortened.

All constituent grasps are activated in parallel, and the first one in time replying with generalized `fail` terminates the whole construct immediately, giving it resultant state `fail` in the SP. Otherwise all branches contribute to the RSP on the rule with values and states associated with them.

5.5.5   Repetitive Branching

For this kind of branching, each new branch starting from the same SP represents another iteration of the same grasp, performed only after the previous iteration completes. The resultant set of props on the rule, RSP, and its resultant control state, RCS, depend on the semantics of concrete repetitive branching rules engaged (their list comprising `cycle`, `loop`, `sling`, and `whirl`).

5.5.5.1   Cycling Branching

This type of repetitive branching, shown in Fig. 5.20, is written in SGL as follows:

$$cycle(G1)$$

The embraced grasp is being repeated as many times as possible if continues to provide resultant generalized state `thru` or `done`, while stopping if its current iteration results with `fail`. The RSP on the whole rule will unite RSPs of all successful invocations of the gasp (accompanied with their AVs and FS). If no iteration succeeds, the rule terminates with `fail` in the starting prop.

Looping Branching

Shown in Fig. 5.21, this branching can be written as:

Fig. 5.20 Cycling branching

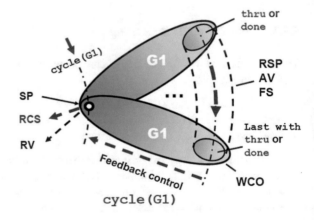

cycle(G1)

Fig. 5.21 Looping branching

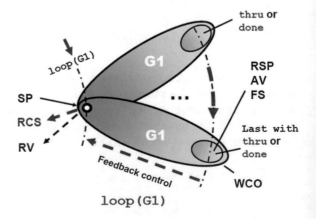

loop(G1)

Fig. 5.22 Slinging branching

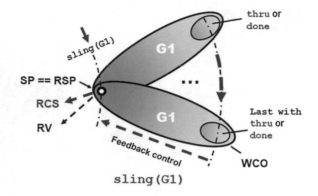

$$loop(G1)$$

This rule develops similarly to the previous one, `cycle`, with the exception that only *last* successful iteration of the grasp (i.e. resulting with `thru` or `done`) contributes to the resulting set of props (with their values and states) of the rule. If no iteration succeeds, the rule terminates in SP with state `fail`.

Slinging Branching

Shown in Fig. 5.22, this type of repetitive branching can be set up as:

$$sling(G1)$$

The rule develops and terminates similarly to the previous rules `cycle` and `loop`, i.e. allowing it to be active until iterations of G1 terminate successfully. But the resulting set of props will always be the SP with state `thru` and its associated latest value, regardless of whether G1 was ever successful or not.

Whirling Branching

This is the last option in the series of repeated branching, called "whirling" (depicted in Fig. 5.23), which can be written as follows:

$$whirl(G1)$$

The rule `whirl` launches the embraced grasp unlimited number of times from the starting prop regardless of success or failure of each its iteration and without producing RSP on the whole rule at all. The iterative invocation of the embraced grasp can be stopped, if needed, by external or special means only, say, by setting a competitive branch which terminates with a success at proper time while killing the

Fig. 5.23 Whirling
branching

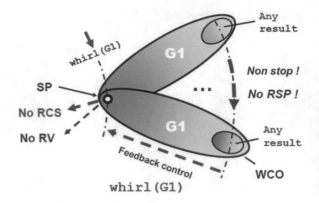

whirling branch still operating (for example, using `or_parallel` rule embracing
the both branches). Another solution may be by providing an additional parameter
explicitly limiting the activity of `whirl` by setting threshold number of its repetitions or the duration time.

5.6   Breadth-Depth Combined Navigation Mode

We have discussed above a number of SGL mechanisms, expressed by its rules, for
integration of different pats of SGL scenarios into more complex space-time navigation structures. This was done separately for depth and breadth modes. Any
combination of these modes can be easily organized in SGL too, which can express
arbitrary parallel and distributed algorithms for solving complex dynamic problems
in large systems and in an integral and compact way. We will consider here only
one example of such breadth-depth integration, with others to be organized in a
similar way.

5.6.1   General Breadth-Depth Mode

We are starting with a very general example of depth-breadth combination of
arbitrary grasps (represented here as G1, G2, G3, and G4) by using general rules
`advance` and `branch`, and expressed as:

$$\text{advance}(\text{branch}(G1, G2, G3), G4)$$

with spatial evolution if this integration shown in Fig. 5.24, where G4 is applied
from RSPs of G1, G2 and G3 developing independently from the same SP.

Fig. 5.24 Combined
breadth-depth evolution of
grasps

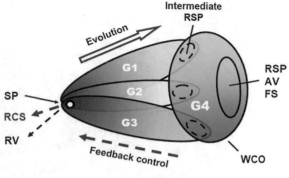

$$\texttt{advance(branch(G1, G2, G3), G4)}$$

Such combination can be further detailed by using such rules as `sequence` and
`parallel`, and modifiers `asynchronous` or `synchronous`. Two cases
detailing this depth-breadth combination are considered below: asynchronous-
parallel and synchronous-parallel.

5.6.2 Asynchronous-Parallel Mode

This can be written as follows:

- `advance_asynchronous(parallel(G1, G2, G3), G4)`
- `advance_async(par(G1, G2, G3), G4)` using shortened rule names
- `advance(par(G1, G2, G3), G4)` same as with `asynchronous`, on
  default.

This combination of parallel branching with asynchronous advancing is shown
at work in Fig. 5.25.

From SP, grasps G1, G2, G3 are developing in parallel, and immediately from
each resultant prop of these grasps, grasp G4 will develop independently, without
any synchronization between different resultant props of each grasp and also
between different grasps. This may drastically simplify the application of SGL code
in a distributed environment, where after splitting into three independent branches
G1, G2, and G3, the grasp G4 can be replicated and attached to each of them
forming three independent scenarios developing in parallel:

- `advance(G1, G4)`
- `advance(G2, G4)`
- `advance(G3, G4)`

where replicated G4 can freely propagate in a distributed space as an attachment to
G1, G2, and G3 during their evolution. Moreover, the already replicated G4 will be
further replicating and developing independently from each prop of RSP1, RSP2,

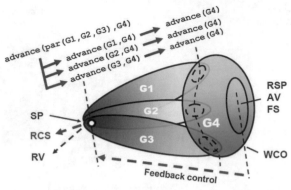

advance_asynchronous(parallel(G1, G2, G3), G4)

Fig. 5.25 Asynchronous-parallel evolution

and RSP3. The feedback control supervised in SP will take place only after full completion of all invocations of G4 in all three branches, and the RSP and AV and FS on the whole breadth-depth composition will be assembled from all final props produced by all copies of G4.

5.6.3 Synchronous-Parallel Mode

This case, depicted in Fig. 5.26, can be represented textually as follows:

- advance_synchronous(parallel(G1, G2, G3), G4)
- advance_sync(par(G1, G2, G3), G4) in a shortened way.

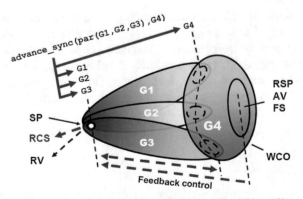

advance_synchronous(parallel(G1, G2, G3), G4)

Fig. 5.26 Synchronous-parallel evolution

In this combined breadth-depth scenario, only after full termination of G1, G2, and G3 activated in parallel from SP, the grasp G4 will be activated from resultant props of all three grasps, providing final props on the rule which, integrating, produce its RSP. This scenario will have two stages of global control (instead of one as in the previous scenario): first after termination of the first three grasps operating in parallel, confirming this termination and allowing the activation of G4; and the second one confirming termination of all invocations of G4 and readiness of final values. G4 can be either propagated as an attachment to G1-G3 and only activated after their full termination by the command from SP, or be broadcast together with this activation command from SP to the resultant props of G1-G3.

5.7   Direct Operations on Remote Values

Due to globally recursive organization of SGL, it is possible to write in it arbitrary complex expressions directly operating not only with local but also any remote data, where any grasps (ranging from just a constant or variable to complex distributed scenarios covering arbitrary areas in physical or virtual worlds) can serve as operands in any operations expressed by rules.

An example of a seemingly traditional data processing & assignment expression but operating on arbitrary complex scenarios (represented by grasps G1, G2, G3, and G4) instead of traditional constants and variables may look like follows:

$$\text{assign}(G4, \text{add}(G1, G2, G3))$$

or in a traditional notation allowed in SGL implementations too:

$$G4 = G1 + G2 + G3$$

A possible evolution and execution of this scenario in a distributed space is shown in Fig. 5.27.

Fig. 5.27 Direct operations on remote data

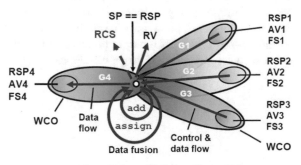

$$\text{assign}(G4, \text{add}(G1, G2, G3))$$

Starting in SP, the grasps G1 to G4 can be launched in parallel. After reaching the final props by independent grasps G1, G2, and G3 (RSP1 to RSP3 in Fig. 5.27) the values associated with these props (AV1 to AV3) are lifted and echoed to SP for their processing (adding to each other). The obtained result in SP (a single number or a list of them, depending on AV1 to AV3, which may be lists) is then sent to the RSP of G4 (RSP4 in Fig. 5.27, which may point at remote variables or node contents to be updated). G4, however, can still be operating toward achieving its RSP4, so the transmission to the latter may have to be delayed until its completion. The RSP on this spatial addition-assignment scenario will include the SP only (usually with state thru) with the associated resultant value of the add operation.

This scenario may have different implementations depending on the operational environment. For example, if we are working with a large distributed space, application of G1 to G4 may involve movement of scenario code to distributed data via a computer network, and then (for G1 to G3) movement of the obtained data back through this network. In case of G4, the result of addition of G1, G2 and G3 will have to move to the RSP of G4 during or after its evolvement in space. G4 can develop in parallel or ahead of G1 to G3, with its developed paths subsequently used to transfer the resultant data to RSP4. If values obtained by G1 to G3 represent physical matter or physical objects (for example, sand, cement, and water to produce concrete by adding to each other), the processes of obtaining them may involve physical movement to related world locations and then back to SP's physical location to be processed there, and subsequently, as final product, to physical location(s) defined by G4.

5.8   Conclusion

We have considered only some elementary and very general schemes of structuring of spatial SGL scenarios while omitting details of their semantics (like sense of the operations shown, use of spatial variables or nature of navigated environments). All these will be clarified and explained in detail within numerous application examples throughout the rest of this book. The exhibited style of direct parallel multidimensional programming in large distributed spaces can naturally shrink to conventional programming if, for example, the whole scenario should stay and operate in the same world point only. And this will automatically result it traditional *parallel programming* if multiple props are allowed at the same time or traditional *sequential programming* if props can appear only in a sequence, one at a time.

References

1. P.S. Sapaty, *Logic Flow in Active Data in VLSI for Artificial Intelligence and Neural Networks,* ed. By W.R. Moore, J. Delgado-Frias (Plenum Press, New York and London, 1991)
2. P.S. Sapaty, *Mobile Processing in Distributed and Open Environments* (Wiley, New York, 1999), 436 p. ISBN: 0471195723
3. P.S. Sapaty, High-level spatial scenarios in WAVE, in *Proceedings of the International Symposium AROB 5th*, Oita, Japan, Jan 2000, pp. 301–304
4. P.S. Sapaty, *Spatial Programming of Distributed Dynamic Worlds in WAVE*. Presentation at the special colloquium "Internet Challenges", Hasso-Plattner-Institut, University of Potsdam, Berlin, Germany, 4 Oct 2002, 50 p
5. P.S. Sapaty, *Ruling Distributed Dynamic Worlds* (Wiley, New York, 2005)
6. P.S. Sapaty, Distributed Missile Defence with Spatial Grasp Technology. SMi's Military Space, Holiday Inn Regents Park London, 4–5 Mar 2015
7. P.S. Sapaty, Gestalt-based integrity of distributed networked systems. SPIE Europe Security + Defence, bcc Berliner Congress Centre, Berlin, Germany (2009)
8. P.S. Sapaty, Grasping spatial solutions in distributed dynamic worlds. Int. J. Comput. Technol. (IJCT) **3**(4) (2012)
9. P.S. Sapaty, Integration of ISR with Advanced Command and Control for Critical Mission Applications, in *SMi's ISR Conference*, Holiday Inn Regents Park, London, 7–8 Apr 2014
10. P.S. Sapaty, High-level Organisation and Management of Directed Energy Systems, Directed Energy Weapons 2010, 25–26 March 2010, CCT, Canary Wharf, London
11. P.S. Sapaty, Cooperative Conquest of Distributed Worlds in WAVE, in *Proceedings of the Symposium and Exhibition of the Unmanned Systems of the New Millennium*, AUVSI'99, Baltimore, MD, 13–15 July, 1999, 16 p

Chapter 6
SGL Networked Interpreter

6.1 Introduction

The chapter reveals general structure and organization of SGL interpreter which may be networked with other similar interpreters for solving distributed problems in parallel and distributed way. Main data structures of the interpreter, its functional processors and their interactions are described which themselves may work together as an integral and parallel system. The scenario history based *distributed track system* of the distributed interpreter allows it to serve as a spatial support and control backbone providing overall awareness, delivering new code and data to the props reached previously, also maintaining and updating distributed information resources.

SGL interpreter can interact with other interpreter copies by any communication channels, where the whole network of them can be mobile and open, capable of changing the number of nodes, their communication structure, also relative and absolute positions in physical space. This interpretation network can effectively implement and support global and local data structures, situation awareness, and overall system control by direct interpretation of evolving spatial SGL scenarios in parallel and distributed manner.

The initial structure of such interpreter for the SGL's predecessor language called WAVE [1–5] was granted a European patent [6]. The previous interpreter versions were also implemented in different countries in Lisp and C (both compiled and interpreted variants) and were used for different purposes, especially for intelligent network management, distributed interactive simulation, virtual reality systems, and mobile cooperative robotics [7–20]; they were also in public domain.

© Springer International Publishing AG 2017
P.S. Sapaty, *Managing Distributed Dynamic Systems with Spatial Grasp Technology*, Studies in Computational Intelligence 690,
DOI 10.1007/978-3-319-50461-2_6

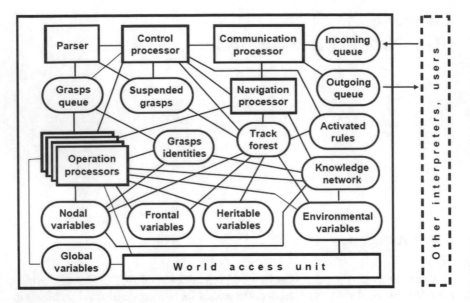

Fig. 6.1 SGL interpreter organization and main components

6.2   The Interpreter General Organization

The SGL interpreter general organization and its main components are shown in Fig. 6.1.

The interpreter consists of a number of specialized functional processors (shown by rectangles) working with and sharing specific data structures. These include: Communication Processor (CP), Control Processor (COP), Navigation Processor (NP), Parser (P), different Operation Processors (OP), and special (external and internal) World Access Unit (WAU) directly manageable from SGL. Main data structures (also referred to as *stores*) with which these processors operate (shown by ovals) comprise: Grasps Queue (GQ), Suspended Grasps (SG), Track Forest (TF), Activated Rules (AR), Knowledge Network (NN), Grasps Identities (GI), Heritable Variables (HV), Fontal Variables (FV), Nodal Variables (NV), Environmental Variables (EV), Global Variables (GV), Incoming Queue (IQ), and Outgoing Queue (OQ).

6.3   Data Structures of the Interpreter

As SGL interpretation network generally serves multiple scenarios or their parallel branches simultaneously navigating the distributed world, which can cooperate or compete with each other. Each interpreter is supporting and processing multiple

SGL scenario code that happens to be in its responsibility at different moments of time. This is reflected in the interpreter's basic data structures which also hold permanent, persistent, and temporary information about the navigated and processed worlds.

6.3.1 Grasps Queue

The Grasps Queue (GQ) keeps multiple scenario fragments (syntactically represented as *grasps*) which are parsed and ready for execution in this particular interpreter. Independent from each other and queued by the priority or arrival time, they can be processed sequentially or in parallel (if the latter capabilities exist in the interpreter). Partially executed grasps (say, needing additional execution functionalities) can be returned to the same GQ for further processing.

6.3.2 Suspended Grasps

Suspended Grasps (SG) structure keeps deferred parts of SGL scenarios which will be parsed in detail and executed in this interpreter or forwarded to other interpreters when proper conditions are met, control states and/or data availability including. The sequence of recorded grasps in SG does not represent a queue, as the time and order of their invocation depend on situations within distributed control of multiple operations in the current and other interpreters.

6.3.3 Track Forest

Track Forest (TF) is a special self-optimizing structure reflecting dynamic history of spatial evolution of SGL scenarios, allowing us to automatically coordinate and control multiple distributed processes with making decisions at different levels. It preserves integrity of the whole set of parallel and distributed processes as a global goal-driven system, also supports existence and controls lifetime of different types of spatial variables described in Chaps. 3 and 4. TF spans throughout the navigated world with its interlinked parts kept in different interpreters while forming altogether a seamless spatial organizational, command and control infrastructure. Its work will be detailed later in this chapter.

6.3.4 Activated Rules

Activated Rules (AR) store represents rules that have been activated and continue working with different scenarios or their branches. Some rules may be waiting for their operands to be completed (say, for obtaining data to be processed or the echoed control states to make decisions). These operands may be arbitrary grasps evolving in the same or in other interpreters and integrated within SGL scenarios by means of track trees in TF. Rules registered in AR are associated with certain TF nodes and the latter, in their turn, with related grasps in GQ and SG which may represent the rule's operand scenarios currently in active, passive, or suspended states.

The rules in AR may be of any kinds—from simple ones to their aggregates with personal parameters, which may need to be processed first to obtain the values needed. The data-processing rules with all operands completed (i.e. which can be directly executed) are being placed into GQ. Rules in AR may form a hierarchy covering via tracks more than one interpreter and operating as an integral system with the help of distributed TF structures.

6.3.5 Knowledge Network

The Knowledge Network (KN) store keeps interlinked data on parts of virtual, physical, execution, or combined worlds associated with the current interpreter, this including named or unnamed nodes (which may have physical coordinates) and named links connecting them. These parts may belong to different applications with their specific world domains, which are navigated and processed by different SGL scenarios, where the latter may cooperate, compete, or just ignore each other.

Nodes of KN fragments in the current interpreter may have semantic links with nodes located in other interpreters thus forming integral distributed information infrastructures covering application areas. Any node navigated or created by the current interpreter and belonging to PW, VW, EW or their combinations is immediately registered in the KN store. Except unnamed pure PW nodes, which automatically disappear as soon as all activities related to them cease, all other nodes and relations between them in KN remain regardless of being or not being accessed by SGL scenarios, unless explicitly deleted by special rules.

6.3.6 Grasps Identities

The Grasps Identities (GI) are keeping personal colors, or identities, of scenarios that have been currently processed in this interpreter (in an active or suspended way), directing them to the resources related to these scenarios, like nodal variables and KN elements. These can be accessed and shared by different scenarios or their branches with the same identities (or which are aware of these identities while being themselves of other personal colors). This coloring of distributed resources (potentially hierarchical) can be effectively used for protecting own information from unauthorized access while allowing users to work cooperatively with other users on distributed resources from different application areas.

6.3.7   Heritable Variables

The Heritable Variables (HV) store keeps all heritable variables with their contents created by SGL scenarios within the current interpreter, these variables being linked to the Track Forest nodes within which they originated. The heritable variables can be accessed by all processes stemming from these track nodes, which may be developing in this or other interpreters. This access is achievable via propagation through the distributed track trees (and between interpreters, if needed) in both read and write mode.

6.3.8   Fontal Variables

The Frontal Variables (FV) store holds all frontal variables with their contents registered at the current moment of time within the current interpreter. These variables are linked with the fringe nodes of evolving track trees reflecting space-time evolution of generally hierarchical and parallel processes, being always associated with the latest props of the interpreted scenarios (having moved to them from the preceding props). Frontal variables are a sole property of these latest scenario stages; they will be replicated if the scenario splits into branches, becoming a non-sharable property of each new branch. If scenarios have to move to other interpreters, frontal variables will be moving with them too, withdrawing themselves from FV store in the current interpreter.

6.3.9   Nodal Variables

Nodal Variables (NV) store keeps all nodal variables with their contents which are linked to nodes in the KN store being their temporary property until the scenarios that created them remain alive in the distributed space (not necessarily in the current interpreter only). For that reason, they are also connected with corresponding nodes in TF store, as after the full completion of scenarios all their track structures, which may spread between interpreters, automatically disappear together with heritable and nodal variables linked to them. Nodal variables are also connected to grasps identities in the GI store through which they can be accessed and shared by all SGL scenarios having these identities (or knowing them).

6.3.10   Environmental Variables

Environmental Variables (EV) store keeps special variables with reserved names, which allow us to control and work directly with different space and time features of the worlds created, navigated, and processed, also with internal parameters of the interpreter. Most of these variables instead of having their own contents (like the previous heritable, frontal, and nodal ones) are referring to parameters of other structures or the worlds currently navigated (like those registered in KN) by evolving SGL scenarios. They may also be accessing special hardware or software represented by the World Access Unit (like different timing devices, sensors, and channels), or allowing us to directly communicate with local human users or various external devices. Most of these variables are classified as stationary but some may behave like frontal ones (e.g. scenario colors) temporarily linking with nodes in TF store on their spatial move.

6.3.11   Global Variables

Global Variables (GV) store keeps information on global variables, being the most expensive ones, which can be simultaneously used and shared by any scenarios or their different branches, and at any time. Their management is beyond traditional distributed track-supported interpreter organization and can be implemented with the help of Word Access Unit capabilities allowing for direct access of global external stores (or other systems) keeping their contents. In many cases, however, the effect of global variables can be achieved by the use of heritable variables if the latter are declared ahead of the scenario development, but only for individual scenarios, not sets of them where they are independent from each other.

6.3.12   Incoming Queue

Incoming Queue (IQ) accepts and stores incoming messages, which may be complete SGL scenarios from users or their parts coming from other interpreters, to be executed in the current interpreter and, possibly, continued in other interpreters afterwards. The messages may also be remotely obtained control states or data to be analyzed and processed in this interpreter or forwarded further. The incoming messages may have different priorities, with control messages being of highest urgency, so the procedures for processing IQ elements must take into account their both arrival order and importance.

6.3.13   Outgoing Queue

The Outgoing Queue (OQ) store accumulates obtained results to be passed to the directly connected users, parts of SGL scenarios with accompanying intermediate data to be forwarded to other interpreters for further consideration, control states and data to be returned to other interpreters on their request, as well as direct commands to users or neighboring interpreters. As for the IQ, the procedures of processing OQ messages may combine the first-come-first-served strategy with superiority of command, control and emergency messages exchanged with other parts of the distributed system.

6.4   Functional Processors

Functional processors carry different SGL interpretation loads, serve interpreter's data structures, and communicate with other interpreters and external users. They also provide system integrity and overall command and control of local and global operations in potentially distributed and dynamic environments.

6.4.1   Communication Processor

Communication Processor (CP) receives SGL scenarios or their parts, service and control messages, and requested remote data from the external world or other interpreters, classifying them accordingly and sending to other processors for execution. CP also allows the current interpreter to serve as a transit node in exchanges between other interpreters. The incoming and outgoing messages are optimized by CP for efficient communication with other interpreters, also for processing within the current interpreter. CP directly operates with IQ and OQ, and communicates with Control Processor and Navigation Processor via internal buffers, allowing them to operate asynchronously and in parallel with each other.

6.4.2   Parser

Parser (P) carries out syntactic analysis of the scenarios or their parts, extracts control rules in them, decomposes SGL strings into ready for execution elementary grasps (with clarified operands) and the remainders to be suspended for further consideration. The latter can take place when proper conditions are met or values of operands for rules (which themselves may be arbitrary grasps) are finally obtained. Parser also optimizes and compresses SGL code for its further processing (like removing blanks and substituting rule names and special words by short abbreviations). Parser directly operates with GQ and SG, also communicates with Control Processor via internal buffer allowing the two processors to work asynchronously and in parallel with each other.

6.4.3   Operation Processors

Operation Processors (OP) unit performs basic analysis and operation procedures over information units and physical matter (or physical objects), expressed by rules. It works directly with GQ, GI, NV, FV, HV, EV, GV, and KN stores. OP also directly communicates with Control Processor, Navigation Processor, and World Access Unit. In case of compound rules, the rest of SGL string taken from GQ for an operation and partially processed in OP can be returned to GQ for a continued execution by other operation processors.

6.4.4   Navigation Processor

Navigation Processor (NP) specializes in creation and navigation of network-structured data, operating directly with TF and KN stores. These data networks (more persistent in KN and temporary in TF) may both be distributed between interpreters. In this case NP performs network navigation in the current interpreter while transferring control and orders to NPs in other interpreters when reaching the network boundaries in the current interpreter, for a continued navigation. As regards the history tracks in TF, they may be optimized (by substituting node sets with single nodes when history details become redundant) or removed partially or completely by NP upon termination of SGL scenarios or their parts. NP cooperates directly with Control Processor, Communication Processor, and Operation Processors.

6.4.5   Control Processor

Control Processor (COP) provides local and global control and coordination of sequential and parallel processes within the interpreter including classification and forwarding of different types of messages within and between interpreters and decomposition of SGL strings while sending them for execution or suspending till proper conditions met. COP also provides interpretation of all control rules of SGL, supporting the distributed command and control hierarchy based on history tracks, which may spread to and cover other interpreters. COP directly operates on GQ, SG, TF, and AR stores. It also cooperates with CP, P, NP, and OP, supervising collective work of these processors within different SGL interpretation procedures. COP plays the key role in organization and support of track-based distributed management and control.

6.4.6 World Access Unit

World Access Unit (WAU) offers an extension to the interpreter's main functionality for interaction with external (and interpreter's internal) physical and virtual environments, also an interface for integration with other systems like internet, robotic equipment, and all those working with physical matter or objects, global variables too. WAU can be accessed via environmental and global variables as well as directly from processors in OP.

6.5 Track-Based Automatic Command and Control

As both backbone and nerve system of the distributed interpreter, its hierarchical spatial track system dynamically spans the worlds in which SGL scenarios evolve, providing automatic control of multiple distributed processes. Its part related to the current interpreter is kept in Track Forest store which is interlinked with similar parts in other interpreters, forming altogether global control coverage. Self-optimizing in parallel echo processes, this (generally forest-like) distributed track structure provides hierarchical command and control as well as remote data and code access. It also supports spatial variables and merges distributed control states for making decisions at different organizational levels. We consider here some features of this history-based distributed control and management system in more detail.

6.5.1 Track-Based Management Components

The main components engaged in track-based control with their icons used in the subsequent diagrams are shown in Fig. 6.2.

These components with their name types ("i" standing for any number) being as follows:

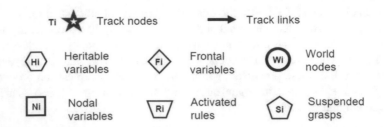

Fig. 6.2 Track-based management components

- *Track nodes* (Ti) reflecting scenario progress points (props)—the stages through which spatial scenarios evolve and form their development history.
- *Track links* showing transition, succession between consecutive props.
- *Heritable variables* (Hi) as sole track node properties capable of being accessed by all operations related to the current and subsequent props.
- *Frontal variables* (Fi) accompanying the scenario evolution and being associated at any time with the latest, or fringe track nodes.
- *World nodes* (Wi) belonging to virtual, physical, executive or combined worlds navigated by the scenario and registered in KN store; these being generally linked with sets of track nodes (as sequences or trees rather than single track nodes).
- *Nodal variables* (Ni) associated with the world nodes created and/or navigated by SGL scenarios. These variables are also linked with particular track nodes under which they were formed. If these track nodes are removed in the tracks cleaning process caused by termination of scenarios or their parts, these variables will be removed too. Nodal variables will also be deleted if the world nodes to which they belong cease to exist (say, after explicit removal in the scenarios using them).
- *Activated rules* (Ri) linked with certain track nodes. They start their influence within corresponding props and may use the subsequent track tree (to its full depth) emanating from these nodes for managing and supervising of the rule-related forward and echo operations.
- *Suspended grasps* (Si) associated with activated rules and connected to the same track nodes as themselves. Si will be subsequently launched by these rules after proper conditions are met, using track infrastructure emanating from this node for their forwarding.

Ti with links in between, Hi, Fi, Wi, Ni, Ri, and Si are located, correspondingly, in TF, HV, FV, KN, NV, AR, and SG stores of the interpreter mentioned above, also shown in Fig. 6.1.

6.5.2 Forward Grasping

In the forward SGL scenario process, the next steps of scenario development can be considered as staying with the same props and track nodes or forming new track nodes for the new props, connected to the previous ones by track links. Except reflecting history of scenario evolution, this growing track tree, as already mentioned, is supporting heritable, nodal and frontal variables as well as activated rules and suspended grasps (all being associated with proper tack nodes) as shown in Fig. 6.3 for some hypothetic SGL scenario.

Track nodes are directly associated with particular world nodes at which they appear (following the SGT ideology where all SGL processes are always linked with certain world points where they take place) and this association is inherited by all subsequent track nodes unless they fall into alliance with other world nodes. The latter, in turn, will be inherited by the subsequent track nodes unless shifting to responsibility of other world nodes, and so on. Thus for node T1 in Fig. 6.3 created when staying with world node W1 the subsequent track nodes T2, T3, and T4 will be associated with W1 too, but not T5. The latter will fall under world node W2 after the scenario changes to this new world position, and the subsequent nodes T6 and T7 will remain with W2 too.

For the scenario in Fig. 6.3, track node T1 is responsible for rule R1 and related to it suspended grasp S1 (say, waiting for the completion of full tree T1–T7, to be

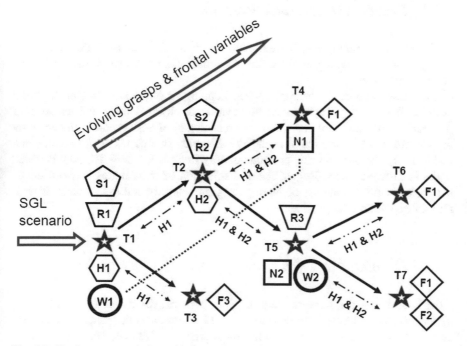

Fig. 6.3 Track system evolution with forward grasping

launched from the final positions reached). It also holds heritable variable H1 which can be accessed from T1 and all subsequent props expressed by corresponding track nodes, i.e. T2–T7, as well as from further grasps if any. Node T2 keeps rule R2 with suspended grasp S2 (the latter, say, may be waiting to be engaged with the second branch T5–T7 after completion of the first branch T4); it also holds heritable variable H2 potentially serving nodes T2, T4–T7, and beyond.

Node T4 (inheriting association with world node W1, commencing with T1) holds and supports life cycle of nodal variable N1 linked to W1 (as its property) which was created within the prop expressed by T4. Node T5 holds and supports life cycle of nodal variable N2 directly linked to world node W2; it also keeps rule R3 used to produce two branches T6 and T7.

And all fringe track nodes T3, T4, T6, and T7 are keeping frontal variables which propagated between track nodes during scenario evolution and reached them at the end of this scenario, where T4, T6, and T7 are holding frontal variable F1 (supposedly created at the prop related to T2). T7 also having F2 (supposedly created at T7), and T3 having F3 (supposedly created at T3). Other frontal variables (not shown in Fig. 6.3) could be brought to this scenario from outside, i.e. via the starting track node T1, reaching the fringe track nodes T3, T4, T6, and T7 after the scenario completion. They will be reunited (along with F1, F2, and F3) with subsequent grasps, if any, and spread further.

6.5.3   Distribution of Track Structure

The track infrastructure can be automatically distributed between different doers during scenario spreading in distributed environments, as shown in Fig. 6.4 for the previous example.

This may happen, for example, when hopping or physically moving to world node W2 from node W1 lying at a certain virtual or physical distance from the latter (and served by a different doer), with the related part of SGL scenario (reflected by track nodes T5–T7) transferred to the W2 area too. This part will retain command and control relations with the previous part by track link T2 → T5, now between nodes in different doers. Regardless of distribution of track nodes between doers, the track system will always be having the same integrity and functionality if were all located within a single doer.

6.5.4   Echoing Via Tracks

After completing the forward stage of SGL scenario discussed above, the track system can return to the starting track node (T1 in our case) the generalized control state based on termination states in all fringe props (i.e. T3, T4, T6, and T7), as in Fig. 6.5, also marking the passed track links with the states returned via them. The

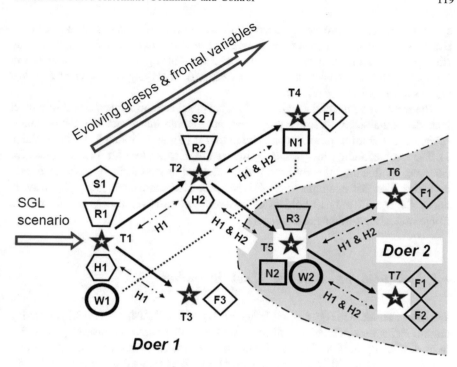

Fig. 6.4 Possible automatic distribution of track infrastructure between doers

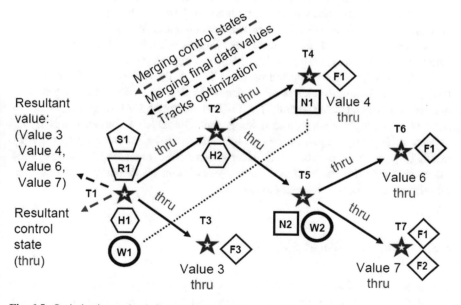

Fig. 6.5 Optimization and echoing and generalization of states via tracks

states generalization process, as described in Chaps. 3 and 4, is based on priority of control states (from strongest to weakest as: *fatal*, *thru*, *done*, and *fail*). The echoing state generalization from the fringe nodes to the starting one will be as shown in Fig. 6.5, with the resultant state *thru* appearing in T1 (taking into account that *thru* was issued in all fringe nodes).

The track system, on the request of higher-level scenario rules, can also collect local data obtained at its fringe props and merge them into a resultant list of values echoed to the starting prop, thus receiving in T1: (Value 3, Value 4, Value 6, Value 7). The track echoing process also optimizes the track system for its further use, for example, by deleting already used and not needed any more items associated with its nodes like, for example, rules R2 and R3, also suspended grasp S2 (leaving nodes T2 and T5 only as holders of variables H2 and N2, also as transit ones to fringe nodes for further grasps, if any).

6.5.5   Failed and Blocked Track Branches

We have considered (Fig. 6.5) all fringe track nodes (T3, T4, T6, and T7) as having control states *thru*, which will all be saved for further use of this track system. But in case of state *fail* in track nodes (fringe ones or intermediate with this generalized state) these nodes will be deleted together with all associated items and track links leading to them. The deletion will not take place (with routes to the starting node saved) if these nodes are holders of nodal variables (the latter must remain till full scenario completion), and the incoming links to them will be marked with *fail*. In case of state *done* in a node, it can echo the value obtained at the related final prop to the starting track node (if requested by a higher rule), but will be denied from a further forward access, with marking the incoming link by *done* too.

So, for example, if *fail* appears in nodes T3 and T4, and *done* in T7 we will be having the remaining track system as shown in Fig. 6.6, with resultant state *thru* and returned value (Values 6 and 7) in T1. The only open for further actions will be node T6 with prop's Value 6, also holding frontal variable F1 that reached it. If *fatal* appears in any track node, the whole track tree will be removed up to the node holding special rule `contain` (if such exists in the preceding track structure), otherwise the entire track infrastructure together with all associated items will be eliminated, and the whole scenario aborted.

6.5.6   Further World Grasping

The echo-modified and optimized track system after the previous scenario stage completion can route further grasps to the world positions reached by the previous grasps and defined by fringe track nodes having state *thru*. This will be twice for

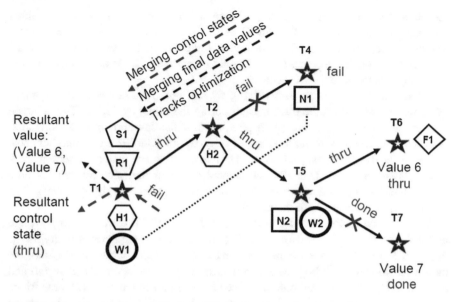

Fig. 6.6 Echo-based tracks modification with failed and completed branches

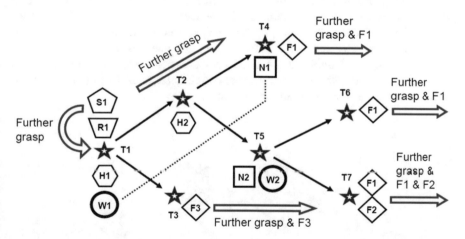

Fig. 6.7 Development of a further grasp

W1 by T3 and T4 and also twice for W2 by T6 and T7, as shown in Fig. 6.7 for the case of Fig. 6.5 with all fringe track nodes having *thru*. The new grasps will be united in these fringe track nodes with frontal variables F1–F3 brought and left there by the preceding scenario stage.

The next scenario stage is supposed to be represented in our example by the suspended grasp S1 associated with rule R1 at the starting track node T1. This rule is believably waiting for the full completion of the previous stage of the scenario, to launch another one from the space positions gained by the previous stage. In other possible variant, instead of R1 with S1 at T1 the new grasp could also be injected via the starting track node T1 from higher organizational levels, where the track fragments considered in Figs. 6.3–6.7 could be part of a larger track infrastructure reflecting a more global SGL scenario.

Copies of the *further grasp* evolving from fringe track nodes of the previous grasp will be extending the track system during their own evolution. The enlarged track system will echo again (with possible modification and optimization) after completion of the new scenario stage, to be ready for transference of another stage if any to its fringe nodes, and so on. Heritable variables created in certain track nodes can be accessed from the subsequent nodes in the track system for both reading and writing, and at any depth of the evolving track tree, by following links between track nodes in both directions.

When spreading the new scenario stage via the track infrastructure to all fringe nodes shown in Fig. 6.7, we assumed that all track links were marked with *thru*, as in Fig. 6.5. But if to have the resultant track infrastructure shown in Fig. 6.6, the new scenario stage will be blocked from spreading via links marked with *fail* and *done*, as in Fig. 6.8 (crossed links). The scenario will reach the only available fringe node T6 with state *thru* (associated with world node W2) and unite with frontal variable F1 already waiting there.

As already mentioned, the sense of saving track node T4 with state *fail* is that it guarantees life cycle of nodal variable N1. The reason of saving node T7 with state *done* is that its status may potentially be changed from higher scenario levels from *done* to *thru* by SGL rule lift, if used. The latter operates in a top-down tracks navigation mode and unblocks the blind alley scenario branches (both nodes and links) which ended with *thru*, thus giving possibility (of node T7 in our case) to be engaged in further forward scenario processes.

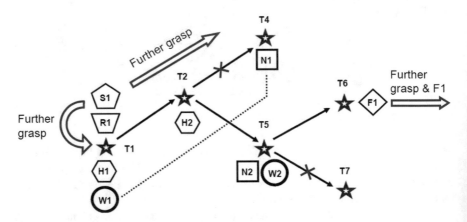

Fig. 6.8 Development of a further grasp through reduced tracking

6.5.7 More Advanced Track Infrastructure

The track system infrastructure for a scenario is usually considered as a history tree (as in Figs. 6.3–6.8) or forest in general case. The latter may be when a single scenario breaks into independent fragments each with own track tree by using rule `release`.

Making track system more advanced than just a tree or forest by adding additional links between track nodes can increase integrity and safety of the distributed system organizations. Thus in Fig. 6.9, in comparison with Fig. 6.7, direct bilateral links are added between some fringe nodes (like T4 and T6, T6 and F7, and T3 and T7, horizontally chaining them) along with direct links from the starting node T1 to all fringe nodes T3, T4, T6, and T7. This can be easily accomplished within track optimization processes, which can be both backward and forward.

It could be even more advantageous to organize a full graph of links between all fringe nodes. Such track system extension (existing together with the previous links between nodes shown by dashed arrows in Fig. 6.9 and supporting scenario-related distributed information infrastructure, including heritable and nodal variables and links to the world nodes) can allow us to reduce time of distribution of further grasps to fringe track nodes with frontal variables from T1. It can also guarantee the reaching of all fringe nodes from the starting node in case of indiscriminate failures, where direct links between fringe nodes could be used for safe distribution of a further grasp between all of them if at least a single one receives it from the starting node, here T1.

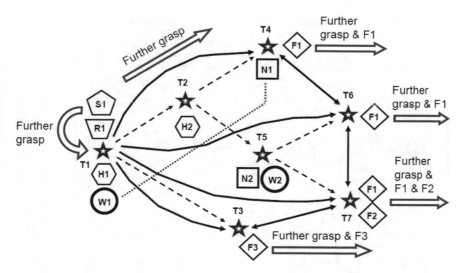

Fig. 6.9 More advanced track structure

6.6   Examples of Involvement of Interpreter Components

We are providing here elementary examples of how different units of the interpreter
participate and interact with each other during execution of SGL code, to give some
impression of the interpreter's functionality, dynamics, and load distribution and
balancing.

6.6.1   Transferring Control Messages

Figure 6.10 shows interpreter's activity (by shaded components) during transference of a control message via it on the way to destination, say, identified by an
absolute physical address. (The current interpreter may have been chosen automatically by the distributed interpretation system as an intermediate communication
stage.) The destination is presumed not to be registered in the current interpreter,
nor have been accessed via it before (this means that history in the Track Forest
cannot be used for routing).

As can be seen, Incoming Queue, Outgoing Queue, and Communication
Processor will be engaged in this operation, where the CP, picking up the message
from IQ, will be calculating the most reasonable next hop (which may be intermediate too) to the destination, and the message will be forwarded to the other
interpreter (or interpreters, in case of a broadcasting transference of control) via the
Outgoing Queue.

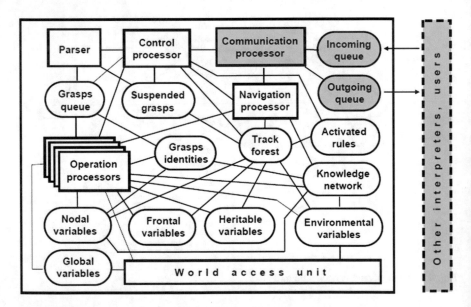

Fig. 6.10 Transference of a control message

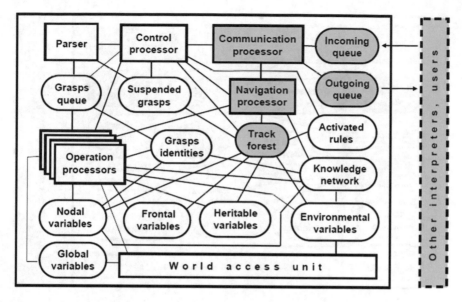

Fig. 6.11 Transference of control and data via track infrastructure

Transferring control or data message via the current interpreter to the destination already visited by the same scenario and located in another interpreter, which can be assisted by traversing the evolution history in TF, is shown in Fig. 6.11. With the employment of Navigation Processor for this, first moving against track links in the current interpreter and then links in other interpreters (engaging their NPs for this), we can reach the needed destinations and deliver "information cargo" to them.

6.6.2   Engagement in Data Processing

Let us consider engagement of the interpreter components for a routine data processing operation with multiple operands, first without assignment of the result, thus leaving it open, as follows:

$$add(5, 6, 7)$$

For this interpreter engagement case, shown in Fig. 6.12, the SGL scenario will be received from the user into IQ by Communication Processor, which will send it to Control Processor. The latter will forward it to Parser to verify SGL syntax, and if correct and the text is directly executable, the Parser will further the scenario via Grasps Queue to Operation Processors to choose the appropriate processor. The latter will perform this addition operation over the three operands, which are ready for execution.

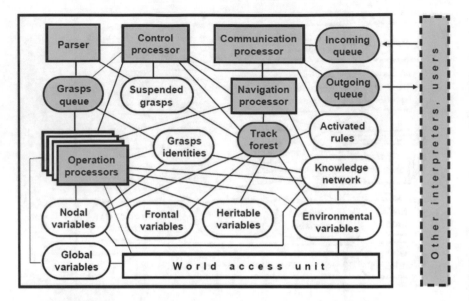

Fig. 6.12 Engagement in data processing operation without assignment

Without assignment of the result, it will be lost after completion of this SGL code, but the operation's termination will be acknowledged to the user first via COP and then by CP using Outgoing Queue. Also, the history of performing this addition operation will be formally registered as a new node in TF using Navigation Processor, this node to be removed after the scenario completion.

Extension of the previous example by assignment of the result to a variable, say nodal, may look like follows:

$$\text{assign}(NN, \text{add}(5,6,7)$$

It will engage more interpreter components, as shown in Fig. 6.13, where such structures as AR, SG, NV, and GI will be additionally used.

While add(5, 6, 7), as before, is recognized as directly executable code to be sent to OP, the rule assign will be recorded in AR and activated before this, with additional control node fixed in TF. The variable name NN will also be recorded in the SG and will be used afterwards together with the recorded rule assign and the result of add after its completion in OP to form the new directly executable code assign(NN, 18) by COP, which will send it via GQ to OP. The latter will perform the assignment to a newly created nodal variable NN, which will be registered in NV store with the color of this SGL scenario (assigned automatically to it by COP from the start) put into GI. The latter would permit further access to this variable only by grasps with the same identity, if more scenario code added to the current one.

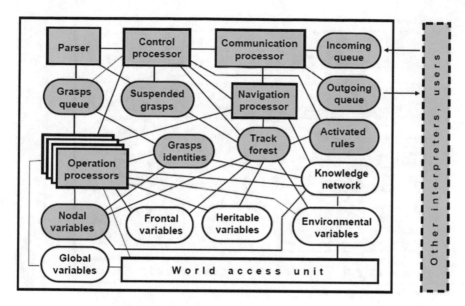

Fig. 6.13 Extending data processing with assignment

6.6.3 Networked Knowledge Processing

Let us consider a simple networked knowledge processing example where, starting at an arbitrary KN node, we should change names of all neighboring nodes reached by links named r to the name of the starting node:

$$\text{hop}(\text{node}(\text{any}));$$
$$F = \text{CONTENT}; \text{hop}(\text{link}(r, \text{all})); \text{CONTENT} = F$$

The frontal variable F is used here to transfer the current node name to neighboring nodes, where node names are accessed by environmental variable CONTENT when staying in them (we could also use another environmental variable NAME here, see Chap. 4). Figure 6.14 shows engagement of the interpreter components for this case, where the starting node (chosen randomly) is supposedly located in the current interpreter and the neighboring nodes r-linked to it may be in the same or other interpreters. For the latter case, with r links connecting network nodes in different interpreters, the engagement of components in other interpreters will be similar to Fig. 6.14.

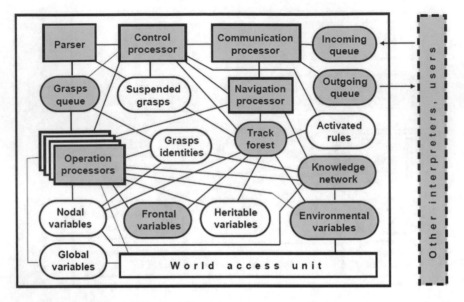

Fig. 6.14 Engagement in networked processing

6.6.4 Movement in Physical Space

Let us consider the case when staying in some physical world position we need to move to another position defined by physical coordinates (x, y) and with proper speed, subsequently sending physical parameters of the position reached to a user. This will be expressed in SGL as follows:

$$\text{move}(\text{coordinates}(x, y), \text{speed}(30));$$
$$\text{OUT} = \text{QUALITIES}$$

This scenario will cause the current interpreter to move to the given physical destination with certain speed and fix the corresponding temporary node in its KN storage. The distributed interpretation network can also appoint another interpreter to physically move to this world location if, for example, the latter happens to be closer to it at the current moment of time. In case of the current interpreter doing the whole job, the engagement of interpreter units will be as shown in Fig. 6.15, where the World Access Unit will provide (directly or serving as an interface to special external equipment) movement to the needed world location and access to its parameters under the request from environmental variable QUALITIES, with subsequent reply to a local user by assigning the obtained result to environmental variable OUT.

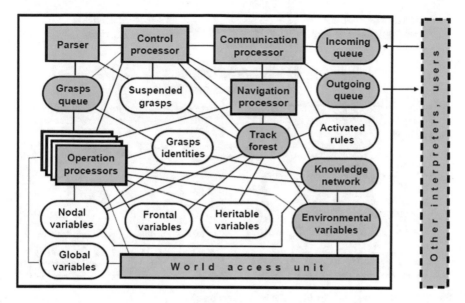

Fig. 6.15 Engagement with movement in physical space

We have shown here only simplest examples of SGL code with different interpreter components engagement and cooperation. For more complex scenarios all interpreter components may be engaged and work in parallel, as well as different, distributed and networked interpreter copies between themselves.

6.7 Integration with Other Systems

Installing SGL interpreters as universal control modules U into other systems like mobile robots, on top of the existing functionality (as shown in Fig. 6.16) allows us to obtain effective group solutions of complex problems in distributed physical spaces (incl. any swarming and advanced global-goal-oriented group behaviors), effectively shifting traditional management routines to automatic levels. Human-robot interaction and integration with gradual transition to fully unmanned systems can be essentially simplified too.

Examples of possible grouping of different robotic units under SGT scenarios into integral teams pursuing global goals and operating under unified command and control are shown in Fig. 6.17.

The dynamically networked SGL interpreters extended by and integrated with other facilities and gadgets, like mobile robots above, can form *universal spatial machines* operating with both information and physical matter, as symbolically shown in Fig. 6.18. These networked machines, working without any central resources under intelligent scenarios injected at any time and from any nodes, can

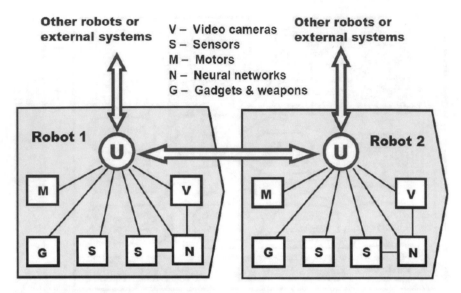

Fig. 6.16 Integrating with traditional robotic functionality

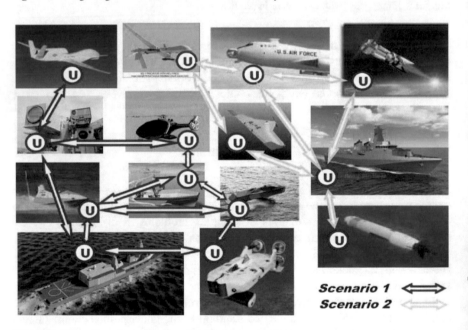

Fig. 6.17 Examples of possible heterogeneous unmanned teams

perform complex computational, knowledge processing and control operations, while moving partially or as a whole in both physical and virtual environments, runtime changing their components, shapes, and coverage area and effectively responding to unpredictable and symmetric situations.

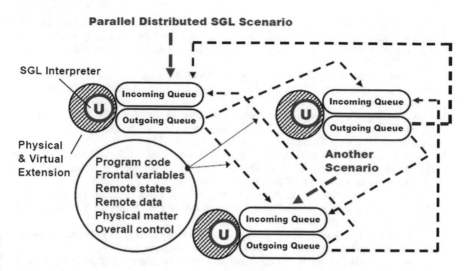

Fig. 6.18 SGL interpretation network as a universal spatial machine

6.8   Conclusions

We have described the organization of SGL interpreter as a basic element of a universal spatial machine which may have any topology and any area coverage, and can operate in an integral and holistic way under ubiquitous self-spreading scenarios in a special high-level language. Based on the networks of SGL interpreters, these machines can comprise any number of nodes represented by humans with laptops, smart phones or smart watches, internet hosts, mobile robots, smart sensors, any other machines, gadgets and devices that can be potentially controlled electronically. The distributed dynamic systems empowered with such spatial machines can be effectively organized for solving any world problems—from social to economic to infrastructure protection to defense—in a holistic and global goal driven way. Many such examples will be considered in the subsequent chapters.

References

1. P.S. Sapaty, A wave language for parallel processing of semantic networks. Comput. Artif. Intell. **5**(4) (1986)
2. P.S. Sapaty, I. Kocis, A parallel network wave machine, in Proc. 3rd Int. Workshop PARCELLA'86 (Akademie-Verlag, Berlin, 1986)
3. P.S. Sapaty, S. Varbanov, A. Iljenko, The WAVE model and architecture for knowledge processing. Proc. Fourth Int. Conf. Artificial Intelligence and Information-Control Systems of Robots (Smolenice, 1987)
4. P.S. Sapaty, The WAVE machine project. Proc. IFIP Workshop on Silicon Architectures for Neural Nets, St. Paul de Vence, France, 28–30 Nov

5. P.S. Sapaty, W. Zorn, The WAVE model for parallel processing and its application to computer network management. Intl. Networking Conference INET'91 (Copenhagen, 1991)
6. P.S. Sapaty, A distributed processing system. European Patent No. 0389655 (European Patent Office, 1993)
7. P.S. Sapaty, P.M. Borst, An overview of the WAVE language and system for distributed processing in open networks. Technical Report, Dept. Electronic & Electrical Eng, University of Surrey, 1994
8. J.C.C. Darling, P.S. Sapaty, M.J. Underhill, Distributed virtual reality: a fully dynamic approach. Proc. 15th Workshop on Standards for the Interoperability of Distributed Simulations (IST UCF, Orlando, FL, Sept 1995), pp. 123–133
9. P.S. Sapaty, P.M. Borst, M.J. Corbin, J. Darling, Towards the intelligent infrastructures for distributed federations. Proc. 13th Workshop on Standards for the Interoperability of Distributed Simulations (IST UCF, Orlando, FL, Sept 1995), pp. 351–366
10. P.S. Sapaty, M.J. Corbin, S. Seidensticker, Mobile intelligence in distributed Simulations. Proc. 14th Workshop on Standards for the Interoperability of Distributed Simulations (IST UCF, Orlando, FL, Mar 1995), pp. 1045–1058
11. J.C.C. Darling, P.S. Sapaty, Distributed dynamic virtual reality in WAVE. In Proc. European Simulation Symposium (ESS-96) (Genoa, Italy, Oct 1996), pp. 36–40
12. P. Sapaty, *Mobile Processing in Distributed and Open Environments* (John Wiley & Sons, New York, 1999)
13. P. Sapaty, M. Sugisaka, Distributed artificial brain for collectively behaving mobile robots. Proc. Symposium & Exhibition Unmanned Systems 2001 (Baltimore, MD, 31 Jul–2 Aug), 18 p
14. P. Sapaty, M. Sugisaka, Towards the distributed brain for collectively behaving robots. Proc. International Conference on Control, Automation and Systems, ICCAS 2001 (Cheju National University, Jeju Island, Korea, 17–21 Oct 2001), pp. 571–574
15. P.S. Sapaty, M. Sugisaka, A language for programming distributed multi-robot systems. Proc. of The Seventh International Symposium on Artificial Life and Robotics (AROB 7th '02) (B-Com Plaza, Beppu, Oita, Japan, 16–18 Jan 2002), pp. 586–589
16. P. Sapaty, M. Sugisaka, Universal distributed brain for mobile multi-robot systems, in *The book "Distributed Autonomous Robotic Systems* ed. by H. Asama, T. Arai, T. Fukuda, T. Hasegava (Springer-Verlag, Tokyo, 2002), SPIN: 10869189, pp. 434–443
17. P. Sapaty, M. Sugisaka, Optimized space search by distributed robotic teams. Proc. International Symposium on Artificial Life and Robotics (AROB 8th) (Beppu, Japan, 24–26 Jan 2003), pp. 189–193
18. P. Sapaty, K. Kawamura, M. Sugisaka, R. Finkelstein, Towards fully distributed cognitive systems. Proc. Ninth International Symposium on Artificial life and Robotics (AROB 9th) (Beppu, Japan, Jan 2004), pp. 274–278
19. P. Sapaty, V. Klimenko, M. Sugisaka, Dynamic air traffic management using distributed brain concept. Proc. Ninth International Symposium on Artificial Life and Robotics (AROB 9th) (Beppu, Japan, Jan 2004), pp. 156–159
20. P.N. Sapaty, M.S. Mirenkov, M. Osano, Distributed artificial life using world processing technology. Proc. of the Fifth Int. Conference on Human and Computer (HC-2004) (The University of Aizu, Japan, 1–3 Sept 2004), pp. 24–29

Chapter 7
Creation, Activation, and Management of a Distributed World

7.1 Introduction

The chapter shows how a hypothetical distributed network-shaped world can be created from scratch, modified, and managed in SGL, in order to demonstrate the peculiarities of SGL and its advanced capabilities of working with interlinked distributed infrastructures.

Different shape-forming distributed methodologies are presented which include those based on depth and breadth first spanning trees, creating and interconnecting networks by parts, also distributing them in physical space. The created distributed world is then inhabited by active mobile agents that can freely migrate between network nodes, see each other, and report to an outside center. Activity will also be added to the communicating network nodes which can interact with active objects moving through the network and inform other nodes of the situations observed.

Additional higher layers can be easily added to the created world which can provide local and global awareness of the distributed situations and impact the overall world behavior. The higher layers can also support an outside control and tasking of the distributed system. The created active world can be easily restructured with its parts removed or new ones added at runtime without interruption of the overall activity. It is also shown how the created virtual world can move in distributed physical space in parts or as a whole, preserving its integrity and shape if needed, how it can be converted into a physical world, and also how physical networked structures can be composed from the very beginning.

The chapter's material is influenced by some previous publications in this field, especially [1–14].

© Springer International Publishing AG 2017
P.S. Sapaty, *Managing Distributed Dynamic Systems with Spatial Grasp Technology*, Studies in Computational Intelligence 690,
DOI 10.1007/978-3-319-50461-2_7

7.2   Distributed World Creation

We are starting here with virtual networks, whereas working with physical world will be discussed later in this chapter.

7.2.1   Elementary Examples

Some elementary networks to be created in SGL are shown Fig. 7.1, with the creative SGL scenarios following.

- Creating a single isolated node with name 1, as in Fig. 7.1a:

```
create(node(1))
```

- Simultaneous creation of isolated nodes named 1, 2, and 3, Fig. 7.1b:

```
create(node(1), node(2), node(3))
```

- Now let us add some links between the already created nodes. Connecting existing nodes 1 and 2 with link a, and nodes 2 and 3 with link b, as in Fig. 7.1c, can be done by:

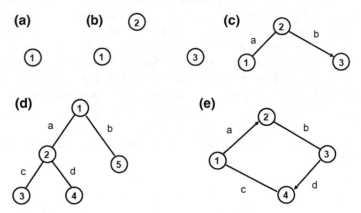

Fig. 7.1 Examples of elementary networks: **a** a single node; **b** a group of isolated nodes; **c** nodes linked in a chain; **d** a tree; **e** a ring

```
advance(hop(node(1)),
     linkup(link(a), node(2)),
     linkup(link(+b), node(3)))
```

Or with just using semicolon to separate consecutive stages instead of rule advance and omitting identification rules link and node if position and use of the given names is clear:

```
hop(1); linkup(a, 2); linkup(b, 3)
```

- Creating network of Fig. 7.1c from scratch starting from node 2 with possibilities of creating nodes 1 and 3 in parallel (using rule create as a global context and comma for separation of independent branches) can be done by:

```
create(hop(2); hop(a, 1), hop(+b, 2))
```

- Creating the tree-like network of Fig. 7.1d starting from its root node 1 and using semicolon and comma instead of rules advance and branch, also parentheses for structuring (with comma superior as operation to semicolon) can be achieved by:

```
create(hop(1); (hop(a, 2); hop(c, 3), hop(d, 4)),
     hop(b, 5))
```

- Creating the ring structure of Fig. 7.1e can be done by:

```
create(hop(1); hop(+a, 2); hop(b, 3); hop(+d, 4));
linkup(c, 1)
```

As node 1 used by rule linkup already exists, more efficient would be using its direct address rather than name for connecting to it, say, by lifting and putting the address into a frontal nodal variable, like F1, right after the node creation (otherwise the node will have to be searched among all existing nodes in the system, which can be more expensive):

```
create(hop(1); F1 = ADDRESS; hop(+a, 2);
     hop(b, 3); hop(+d, 4));
linkup(c, F1)
```

We can also substitute rule `linkup` with rule `hop` under general context of rule `create`, as nodes with assigned addresses already exist and need not to be created again, only links to them if mentioned, as follows:

```
create(hop(1); F1 = ADDRESS; hop(+a, 2);
    hop(b, 3); hop(+d, 4); hop(c, F1))
```

7.2.2 Creating General Networks

In SGL we can create any network topologies, localized or distributed between computers in their network. An exemplary virtual network with named nodes and links (the latter for simplicity considered as non-oriented) is shown in Fig. 7.2.

There are numerous possibilities in SGL of how to create arbitrary network structures, as a whole at once or by pieces, and how to distribute them between different doers, choosing the latter automatically or naming explicitly for network parts or even separate nodes.

7.2.2.1 Network Creation Based on Depth-First Spanning Tree

Amongst effective and universal methods for arbitrary network creation is the one using creative template based on Depth-first Spanning Tree, or DST, covering all network nodes, as shown in Fig. 7.3.

The DSP can be built as follows. Starting from some node as a root, it opens a branch via any link from it and evolves to new nodes as far as possible, following

Fig. 7.2 An arbitrary
network to be created

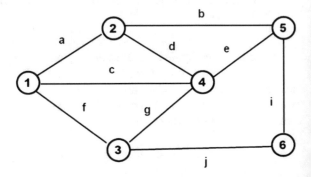

Fig. 7.3 Depth-first spanning
tree of the template

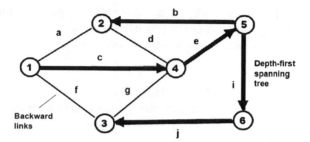

from each node a free link too, and only one, before backtracking to the most
recently passed node still having unused links, to launch another branch, and so on.
For this growing tree, new links from a node can lead only to new nodes or to the
already registered ones lying on the formed path from the root to the current node.

This allows us to effectively organize integral network creation template starting
from any node of the graph and then following the DST formed from this node in a
top-down manner, with branches in nodes, if more than one, capable of being
processed independently and in parallel as they do not share links between them.
Each new node may be assigned a unique address in virtual space or unique
coordinates in physical space (or both in case of VPW), and the addresses and
coordinates of the nodes to which there are backward links in the DST may be fixed
in frontal or heritable variables of SGL, right after the nodes creation, and used
afterwards for forming links to them.

The network creation scenario based on DST of Fig. 7.3 may be as follows (with
names of frontal variables starting with F):

```
create(
  hop(1); F1 = ADDRESS;
  hop(c, 4); F4 = ADDRESS; hop(e, 5));
  (hop(b, 2); hop(a, F1), hop(d, F4)),
  (hop(i, 6); hop(j, 3); hop(f, F1), hop(g, F4)))
```

If to use heritable variables, with their names starting with H, or global ones
starting with G, the scenario will be the same except names of variables, as follows
for heritable variables (more differences may be on implementation level, however).

```
create(
  hop(1); H1 = ADDRESS;
  hop(c, 4); H4 = ADDRESS; hop(e, 5));
  (hop(b, 2); hop(a, H1), hop(d, H4)),
  (hop(i, 6); hop(j, 3); hop(f, H1), hop(g, H4)))
```

Instead of using frontal and heritable variables for keeping addresses of the already passed nodes to which links should be created at further stages, we may just mention these nodes as already passed, as follows:

```
create(
  hop(1); hop(c, 4); hop(e, 5));
  (hop(b, 2); hop_passed(a, 1), hop_passed(d, 4)),
  (hop(i, 6); hop(j, 3); hop_passed(f, 1),
  hop_passed(g, 4)))
```

This scenario can be practically implemented by using internal track system of the distributed interpreter to return to the passed nodes with their names given or by automatically modifying and extending the scenario text with explicit generation and use of frontal, heritable or global variables (as in the previous example) at a preprocessing stage.

A further scenario length reduction can be achieved by using special characters for hop rules with their link and node operands standing on different sides of them. With a hop to new nodes identified by #, and to the already passed nodes by ##, we will have the following scenario text (where direct hop to a node is symbolized by absence of the link operand):

```
create(#1; c#4; e#5; (b#2; a##1, d##4),
              (i#6; j#3; f##1, g##4))
```

This scenario, applied to an empty space, can form the network of Fig. 7.2 starting with node 1 and gradually creating nodes and links connecting them, as shown in Fig. 7.4a–f; it starts in full and then creatively propagates in space losing utilized parts, unless becoming empty. Different DST branches can develop in parallel during the network creation.

7.2.2.2   Creating in Parts to Be Subsequently Interconnected

A variety of other possibilities for creation of distributed structures can be offered in SGL. We may first create parts of a network and then interconnect nodes belonging to different parts by the remaining links between them. In Fig. 7.5, two parts (shown in bold links, one with nodes 1, 2, 5, and another with 4, 3, 6) are to be created first, say, independently and in parallel, and interconnected afterwards (wherever possible in parallel too), which can be accomplished by the following SGL scenario.

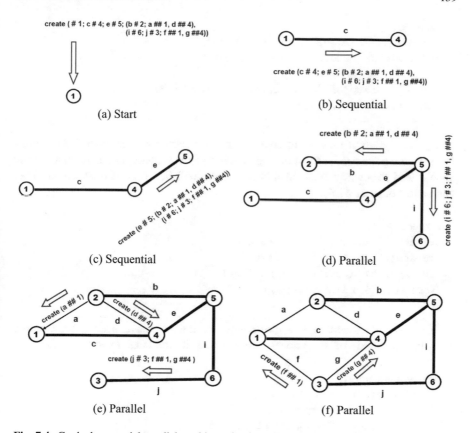

Fig. 7.4 Gradual sequential-parallel world creation by self-evolving and self-modifying template, which loses its worked parts when not needed any more

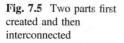

Fig. 7.5 Two parts first created and then interconnected

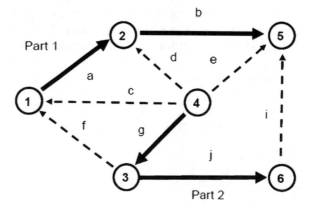

```
sequence(
 ((create(1); create(a, 2); create(b, 5)),
  (create(4); create(g, 3); create(j, 6))),
 ((hop(3); linkup(f, 1)),
  (hop(4); linkup(c, 1), linkup(d, 2),linkup(e, 5)),
  (hop(6); linkup(i, 5))))
```

Another solution may by in creating the two parts in a sequence, where links between the parts can be fixed during creation of the second part, as the nodes they should lead to already exist (in this case there will be no need in repeated direct access to nodes 3, 4, 6):

```
sequence(
 (create(1); create(a, 2); create(b, 5),
 (create(4); linkup(c, 1), linkup(d, 2),
 linkup(e, 5), (create(g, 3); linkup(f, 1),
 (create(j, 6); linkup(j, 5)))))
```

If to use rule `create` as a global context and `hop_existing` for creating links to the already existing nodes, we may rewrite the latest example as follows:

```
create_sequence(
 (hop(1); hop(a, 2); hop(b, 5)),
 (hop(4); hop_existing(c, 1),
 hop_existing(d, 2), hop_existing(e, 5),
 (hop(g, 3); hop_existing(f, 1),
 (hop(j, 6); hop_existing(j, 5)))))
```

The latter may potentially be more expensive than the previously used `hop_-passed`, as finding nodes just by their names anywhere in large distributed spaces may not be easy. Using the abbreviation # for `hop` as before, and ### for `hop_existing`, we may rewrite this scenario shorter, as follows.

```
create_sequence(
 (#1; a#2; b#5),
 (#4; c###1, d###2, e###5,
    (g#3; f###1, (j#6; i###5))))
```

With heritable variables we can fix addresses of nodes in the first part to be used in the second part, by declaring (thus creating) the variables at the beginning of the scenario, as follows:

```
heritable(H1, H2, H5)
create_sequence(
  (#1; H1 = ADDRESS; a#2; H2 = ADDRESS; b#5;
   H5 = ADDRESS),
  (#4; c#H1, d#H2, e#H5, (g#3; f#H1, (j#6; i#H5)))))
```

If to use global variables, there will be no need in their creation in advance, as follows:

```
create_sequence(
  (#1; G1 = ADDRESS; a#2; G2 = ADDRESS; b#5;
   G5 = ADDRESS),
  (#4; c#G1, d#G2, e#G5, (g#3; f#G1, (j#6; i#G5)))))
```

In creative hops to nodes by addresses in the two above cases, usual hop # is used as such nodes already exist and only links to them will be added.

7.2.2.3   Using Breadth First Spanning Tree

Another variant may be based on Breadth First Spanning Tree (BST) covering all nodes (as in Fig. 7.6), with subsequent adding the remaining links to the already existing nodes under `hop_existing` (or `###`).

To guarantee that the latter nodes already exist, we may synchronize the termination of BST before adding the remaining links, as follows (using rule `align` in a nonlocal mode).

Fig. 7.6 Breadth first spanning tree based network creation

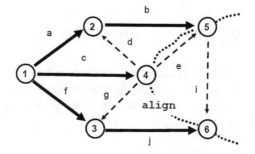

```
create(
  align(#1; (a#2; b#5), c#4,(f#3; j#6));
  (NAME ==5; i###6), (NAME ==4; d###2, g###3, e###5))
```

Or, with prolongation of different braches represented as local procedures in a nodal variable (here N) at fringe nodes after BST termination (which will not need sending all remaining branches into each fringe node and then selecting the proper one):

```
create(
  align(#1;
      (a#2; b#5; N = {i###6}),
      (c#4; N = {d###2, g###3, e###5}),
      (f#3; j#6));
  run(N))
```

With global or heritable variables we can use node addresses when adding the links, similar to the previous cases, as follows for heritable variables.

```
heritable(H2, H3, H5, H6);
create(
  align(#1;
      (a#2; H2 = ADDRESS;
      b#5; H5 = ADDRESS; N = {i#H6}),
      (c#4; N = {d#H2, g#H3, e#H5}),
      (f#3; H3 = ADDRESS; j#6; H6 = ADDRESS));
  run(N))
```

Using different possibilities may depend, for example, on size and topological complexity of the network, peculiarities of implementation, system application areas, etc.

7.3 Network Distribution

The SGL interpreter can distribute network nodes between available computers in different ways, including randomly, as in Fig. 7.7.

Particular computer resources to be used in the distributed creative process can be ordered explicitly, as follows:

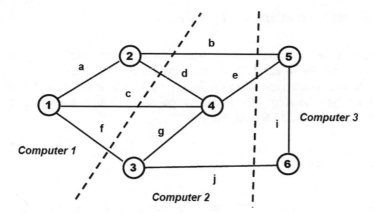

Fig. 7.7 Possible distribution of network nodes

```
RESOURCES = (Computer1, Computer2, Computer3);
create(#1; c#4; e#5; (b#2; a##1, d##4),
            (i#6; j#3; f##1, g##4))
```

We may also appoint a priori an exact computer for each node (using colon as a delimiter to reduce parentheses):

```
create(#1:Computer1; c#4:Computer2; e#5:Computer3;
      (b#2:Computer1; a##1, d##4),
      (i#6:Computer3; j#3:Computer2; f##1, g##4))
```

It is also possible for each node, at any time, to change its residence to another computer, say, for node 5 from the current computer (i.e. Computer3) to Computer1:

```
#5; DOER = Computer1
```

All links of node 5 to other nodes located in the same or other computers will be saved. We could also relocate any groups or all nodes at once to the same computer, as follows for all nodes:

```
#all; DOER = Computer1
```

7.4 World's Invasion with Mobile Objects

Inhabiting the world created above with nameless active mobile objects (let us traditionally call them "agents") which are randomly, freely, and endlessly moving between nodes with certain delays at them (say, 60 s.), as shown in Fig. 7.8, may be done by the following SGL scenario. It begins in parallel in nodes 1, 4, and 5, where agents start their life simultaneously.

```
frontal(Delay = 60); #(1, 4, 5);
free_repeat(sleep(Delay); random#)
```

Using rule `free` together with rule `repeat` cancels accumulation of the history of evolution of this scenario, which may grow unlimited due to endless movement of mobile agents.

Giving personal identity (like names) to these mobile objects (using environmental variable `ID`) and allowing them to see each other at nodes (by self-registering in a shared nodal variable `Stay` at each node upon arrival to it and self-withdrawing own name when departing), while locally reporting the fact of seeing others at this node, may be accomplished by the following scenario (see also Fig. 7.9):

```
frontal(ID, Delay = 60); nodal(Stay);
(ID = Peter; #1), (ID = Simon; #4), (ID = John; #5);
free_repeat(
 if(nonempty(Stay),
   output(ID &' sees '& Stay &' at '& NAME));
 append(Stay, ID); sleep(Delay); remove(Stay, ID);
 random#)
```

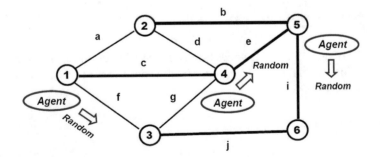

Fig. 7.8 Invading the world with nameless active mobile objects

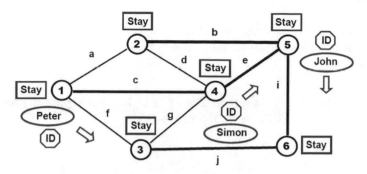

Fig. 7.9 Allowing named mobile objects to see each other at nodes

Under this scenario, for example, when agent Peter arriving at node 4 sees there John and Simon, he will send to the local terminal associated with this node the following:

```
Peter sees (John, Simon) at 4
```

This can also be sent directly to some headquarters (say, represented by node Central, not shown in Fig. 7.9) and output there, as follows:

```
(#Central; OUT) =
  ID & ' sees ' & Stay & ' at '& NAME
```

7.5   Collecting and Exhibiting the History of Navigation

The moving objects we have created can also collect and output information on their passed routes through the distributed networked system, like names of visited nodes and time of entering them as in the following scenario (upon reaching the Limit on the number of records in navigation History the latter is cleaned up, each time):

```
frontal(ID, History, Limit = 100, Delay = 60);
nodal(Stay);
(ID = Peter; #1),(ID = Simon; #4),(ID = John; #5);
free_repeat(
  append(History, NAME & '--' & TIME);
  if(count(History)  == Limit,
    (output(ID &' history: '& History);
    History = nil)));
```

```
if(nonempty(Stay),
  output(ID &' sees '& Stay &' at '& NAME));
append(Stay, ID); sleep(Delay); remove(Stay, ID);
random#)
```

Example of an output in the node where `Limit` has been supposedly reached by moving object `John` (let it be node 6 in Fig. 7.9):

```
John history:
  (5--1501, 2--1503, 1--1505, ..., 6--1640)
```

This can also be forwarded to some `Central` node as explained before.

7.6 Adding Nodal Activity

We can add a permanent personal activity to nodes allowing them, for example, to regularly inform all neighbors (and associated local users) on the objects currently staying at them, like in Fig. 7.10, which may serve as an alarm framework for certain applications.

This may be done by the following SGL scenario additionally injected into all nodes of the already active world (and working with variables already declared during the world creation):

```
#all; nodal(Delay1) = 30;
 sling(nonempty(Stay);
     (all#; OUT) = Stay & ' seen at ' & NAME;
     sleep(Delay1))
```

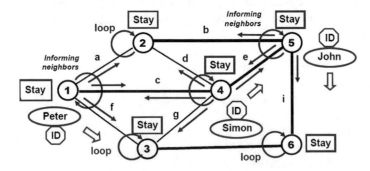

Fig. 7.10 Adding nodal activity informing neighbors on the objects seen

For the situation described in Fig. 7.10 the following messages will be sent with an output at the neighbors:

From node 1 to nodes 2, 3, 4:	Peter seen at 1
From node 4 to nodes 1, 2, 3, 5:	Simon seen at 4
From node 5 to nodes 2, 4, 6:	John seen at 5

We can easily extend this scenario for regular informing not only neighbors but other, more remote nodes too. For example, doing this for all other nodes from each node of the network repeatedly, the following scenario (based on breadth-first spanning tree) can be as follows:

```
#all; nodal(Delay1) = 30; Frontal = Pass;
sling(sleep(Delay1); nonempty(Stay);
    Pass = Stay & ' seen at ' & NAME;
    repeat(all#firstcome; OUT = Pass))
```

This, for example, starting at node 1 for the situation in Fig. 7.10, will be sending to and outputting in all nodes from 2 to 6 the same message (sent only to nearest neighbors in the previous example):

```
Peter seen at 1
```

Each node can also regularly access neighboring nodes and collect data on objects seen there, to be aware of the surrounding on its own initiative, outputting it to a local user, as follows:

```
#all; nodal(Delay2 = 50, Surround);
sling(
 sleep(Delay2);
 nonempty(
  Surround =
   (all#; nonempty(Stay); Stay & '--' & NAME));
 output(
  'Detected at neighbors of' & NAME & ': ' &
  Surround))
```

We will get, for example, from node 3:

```
Detected at neighbors of 3: Peter--1, Simon--4
```

Each node can also regularly see of what appears at other nodes which are not only nearest neighbors, but on any distance from them, all other network nodes including. For the latter case the SGL scenario will be as follows:

```
#all; nodal(Delay2 = 50, Surround);
sling(
 sleep(Delay2);
 nonempty(Surround =
  repeat(
   all#firstcome;
   free(nonempty(Stay); Stay & '--' & NAME)));
 output(
  'Detected at all others from ' & NAME & ': ' &
  Surround))
```

For the same node 3, for example, we will be regularly receiving by this scenario (unless the objects move to other nodes) the following:

```
Detected at all others from 3:
        Peter--1, Simon--4, John--5
```

From node 4 we will be having:

```
Detected at all others from 4: Peter--1, John--5
```

7.7   Global Supervision and Inspection

The active distributed world we have created in the previous sections can be additionally analyzed and managed by injecting other scenarios into it, during its operation and at any time, say, from some external supervising point as shown in Fig. 7.11.

As an example, let us collect all moving objects wherever they happen to be within a single list from this external point and output it there:

```
output_append(#all; Stay)
```

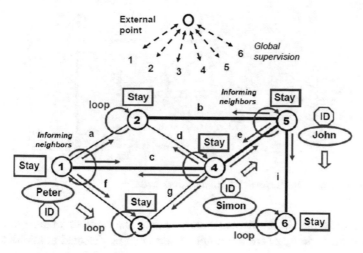

Fig. 7.11 External world supervision

There result will be:

```
Peter, Simon, John
```

We can also change any parameters of the already active networked system, say, Delay1 in all nodes as follows:

```
(#all; Delay1) = 200
```

Another example, finding a node where both Peter and John are currently staying:

```
output(#all; belong((Peter, John), Stay); NAME)
```

Current answer will be: none
Removing Simon from the system wherever this moving object currently stays:

```
#all; remove(Simon, Stay)
```

We can also take any network node as the control center, like 4 as in Fig. 7.12, and access other nodes either directly from it (if supported on the interpreter's level) or by stepwise loop-free navigation of the rest of the network (which may be breadth parallel and tree-like).

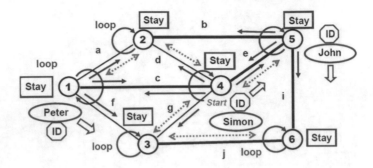

Fig. 7.12 Adding global inspection from a network node

As an example, let us demonstrate counting the number of all other nodes, starting from node 4, where no mobile objects can be detected at this time:

```
output_count(
  #4; repeat(all#firstcome; free_empty(Stay)))
```

It will reply with value 3 for the situation in Fig. 7.12. Similar scenario but for all nodes, the starting node 4 including:

```
output_count(
  #4; repeat(free_empty(Stay); all#firstcome))
```

This will reply with the same value 3 because node 4 is currently occupied with Simon.

7.8   Runtime Restructuring of the Active Distributed World

Any restructuring of this active distributed world can be done at runtime too. For example, removing node 1 with all adjacent links and, independently or even in parallel, adding link w between nodes 2 and 3 and link-node pair (v, 7) from node 6 (as in Fig. 7.13) can be done by the following scenario:

```
delete(1), (#3; linkup(w, 2)), (#6; create(v, 7))
```

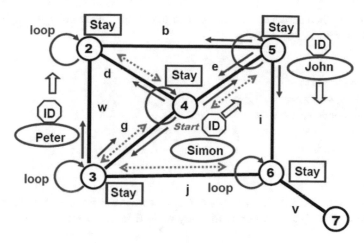

Fig. 7.13 Runtime restructuring of the active distributed world

The modified distributed world will remain active and operational under the changed configuration, same as before. Hopefully, mobile agent Peter was not with node 1 during its removal but, say, with node 3 as in Fig. 7.13 otherwise it could have been deleted from the distributed system together with node 1.

We may delay the removal of node 1 until no mobile objects are residing in it, thus saving all of them during and after the restructuring process—by adding regular checking of its occupation as follows:

```
(#1; loop_if(empty(Stay), delete(CONTENT))),
(#3; linkup(w, 2)), (#6; create(v, 7))
```

The newly created node 7 will not, however, be active as all other nodes, like informing neighboring nodes on the objects it sees, so its creation should be followed by supplying it with this activity (same as previously done for other nodes), so our restructuring scenario will further grow as follows:

```
(#1; loop_if(empty(Stay), delete(CONTENT))),
(#3; linkup(w, 2)),
(#6; create(v, 7); nodal(Stay, Delay1 = 30);
  sling(nonempty(Stay);
     (all#; OUT) = Stay & ' seen at: ' & NAME;
     sleep(Delay1)))
```

Wee may keep this nodal activity from the beginning as a procedure assigned to a variable in each node, and any new node added can simply borrow it from any neighbor and then activate. For this, we can rearrange the previous activity scenario injected into all nodes as follows:

```
#all;
nodal(Activity) = {
  nodal(Stay, Delay1 = 30);
  sling(nonempty(Stay);
      (all#; OUT) = Stay & ' seen at: ' & NAME;
      sleep(Delay1))};
run(Activity)
```

Now, after creation, the new node 7 can borrow the `Activity` content from any neighbor having it, which is the only node 6 in our case:

```
#7;
nodal(Activity) =
  or(all#; nonempty(Activity); Activity);
run(Activity)
```

We could also create the variable `Activity` at node 7 and assign the value to it when we were creating this node from node 6, as follows:

```
(#6; frontal(Transfer) = Activity;
create(v, 7); Activity = Transfer;
run(Activity))
```

7.9   Virtual-Physical World Creation and Management

7.9.1   Converting Virtual to Physical

Let us consider the placement and distribution of the created networked virtual world in physical space too, associating network nodes with certain physical coordinates, as shown in Fig. 7.14.

Placing the already created virtual network into physical world by the given node coordinates can be done by the following scenario:

```
#all; TYPE = (virtual, physical);
(NAME == 2; WHERE = (x_1, y_3)),
(NAME == 3; WHERE = (x_1, y_1)),
(NAME == 4; WHERE = (x_2, y_2)),
```

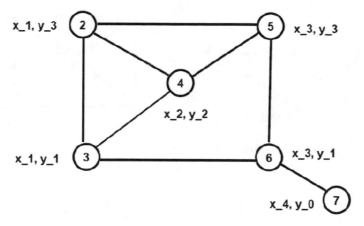

Fig. 7.14 Network nodes with physical coordinates

```
(NAME == 5; WHERE = (x_3, y_3)),
(NAME == 6; WHERE = (x_3, y_1)),
(NAME == 7; WHERE = (x_4, y_0))
```

The statement: TYPE = (virtual, physical) may be omitted as assignment of physical coordinates to virtual nodes will automatically convert these nodes into the combined virtual-physical status.

The scenario above was initially placed into all network nodes, but it is also possible to access each node selectively from the start and place it into proper coordinates, as follows:

```
(#2; WHERE = (x_1, y_3)),
(#3; WHERE = (x_1, y_1)),
(#4; WHERE = (x_2, y_2)),
(#5; WHERE = (x_3, y_3)),
(#6; WHERE = (x_3, y_1)),
(#7; WHERE = (x_4, y_0))
```

7.9.2   Keeping Size and Shape in Movement

We can easily organize movement of this network in physical space, whether in separate parts or as a whole while preserving existing semantic links between different nodes. And this movement may be organized with saving the same physical size and shape of the whole network during the movement, or by allowing

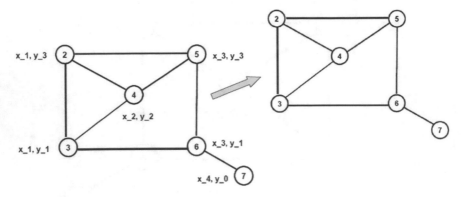

Fig. 7.15 Network movement with its shape and size preservation

nodes move uncoordinated with other nodes, with both size and shape changing in time.

The shape and size preservation can be achieved by the following scenario where each node individually and regularly shifts in physical space in the same direction and distance, also after the same time interval, as in Fig. 7.15.

```
#all; nodal(Delay) = 45;
loop(shift(x_1, y_0.5); sleep(Delay))
```

The movement scenario can be more precise and organized in a globally coordinated and synchronized way, as propagation of separate nodes may happen to differ in speed, say, due to peculiarities of terrain, as follows (where each new iteration will wait for a full completion of the previous iteration step by all nodes):

```
frontal(Delay) = 45;
loop(#all; shift(x_1, y_0.5); sleep(Delay))
```

If we allow a random individual movement of each node (within certain limits), we may have varying network physical size and shape, as shown in Fig. 7.16, which can be organized by the scenario below.

```
#all; Nodal(Delay) = 45;
loop(shift_random(x_(0, 1), y_(0, 0.5));
    sleep(Delay))
```

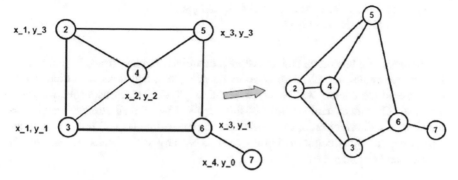

Fig. 7.16 Varying network size and shape with randomized nodes movement

7.9.3 Creating Physical Structures from the Start

Let us create the network of Fig. 7.2 as some physical structure from the very beginning, with nodes associated with locations in physical space and links being physical too, say, as walls or fences with certain height and thickness, connecting these points. We may use the same DSP-based template as before; see Fig. 7.17 and the creative scenario that follows.

```
Fn = "fence: height 2 m, thickness 20 cm";
create(
 #(1, x_1, y_2); F1 = ADDRESS && WHERE;
 (c, Fn)#(4, x_3, y_2); F4 = ADDRESS && WHERE;
 (e, Fn)#(5, x_4, y_3);
   ((b, Fn)#(2, x_2, y_3); (a, Fn)#F1, (d, Fn)#F4),
```

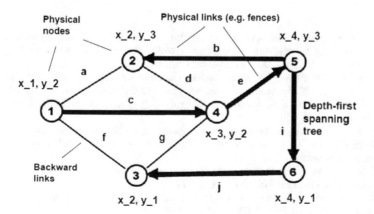

Fig. 7.17 DST-based creation of a physical structure

```
((i, Fn)#(6, x_4, y_1); (j, Fn)#(3, x_2, y_1);
(f, Fn)#F1, (g, Fn)#F4)))
```

All nodes will be automatically assigned combined virtual-physical type, as both their virtual names and physical coordinates are given, and links with their names and physical parameters will be materialized as physical constructs (like fences in our case). Of course, real creation of this physical networked structure may need pursuing of a different order of creation of its physical nodes and links, which can depend on the construction technology used, but any such details can be readily reflected and organized in SGL too.

7.10   Conclusion

The chapter showed how a hypothetical network-shaped world could be created in SGL from scratch and inhabited with mobile objects communicating with each other and with the world. This world can be made active too, self-exhibiting local and global awareness of the situations occurred. It can be restructured at runtime and supervised from outside or inside by setting additional, higher layers. The active world created can be arbitrarily distributed in physical space; it can move through the latter in parts or as a whole while preserving integrity, also acquiring physical qualities itself. The shown examples in SGL, despite very simple and short (can be composed and modified at run time, on the fly), are confirming virtually unlimited potentials of SGT for working with distributed dynamic worlds of any natures.

References

1. P.S. Sapaty, A structural approach to solving tasks of the graph and network theory, in *Proceedings of 5th All-Union School on Parallel Programming and High-Performance Systems*, Kiev, Naukova Dumka, 1982 (in Russian)
2. P.S. Sapaty, On efficient structural implementation of operations on semantic networks, in *Proceedings of USSR Academy of Sciences*. Technical Cybernetics, No. 5, 1983 (in Russian)
3. P.S. Sapaty, A wave language for parallel processing of semantic networks. Comput. Artif. Intell. **5**(4) (1986)
4. P.S. Sapaty, The wave approach to distributed processing of graphs and networks, in *Proceedings of International Working Conference Knowledge and Vision Processing Systems*, Smolenice (1986)
5. P.S. Sapaty, The WAVE-1: a new ideology and language of distributed processing on graphs and networks. Comput. Artif. Intell. **5** (1987)
6. P.S. Sapaty, WAVE-1: a new ideology of parallel processing on graphs and networks. Future Generations Comput. Syst. **4** (1988) (North-Holland)

7. P.S. Sapaty, WAVE: creating dynamic worlds based on mobile cooperative agents. Dartmouth workshop on transportable agents, Dartmouth College, Hanover, New Hampshire, Sept 1996
8. P.S. Sapaty, Distributed modeling of cooperative behavior by mobile agents, in *Proceedings of Sixth Conference on Computer Generated Forces and Behavioral Representation*, IST UCF, Orlando, FL, pp. 599–613 (1996)
9. P.S. Sapaty, Live demonstration of the WAVE system and applications at the workshop on mobile agents and security 97, Maryland Center for Telecommunications Research, Department of Computer Science and Electrical Engineering, UMBC, 27–28 Oct 1997
10. P. Sapaty, *Mobile Processing in Distributed and Open Environments* (Wiley, New York, 1999)
11. P.S. Sapaty, Spatial programming of distributed dynamic worlds in WAVE. Presentation at the special colloquium "Internet Challenges", Hasso-Plattner-Institute, University of Potsdam, Berlin, Germany, 4 Oct 2002, 50p
12. P. Sapaty, K. Kawamura, M. Sugisaka, R. Finkelstein, Towards fully distributed cognitive systems. Math. Mach. Syst., ISSN 1028-9763, No. 1, pp. 69–75 (2004)
13. P. Sapaty, M. Sugisaka, N. Mirenkov, M. Osano, R. Finkelstein, Grasping the distributed entirety, in *Proceedings of Tenth International Symposium on Artificial Life and Robotics* (AROB 10th), Beppu, Japan, 4–6 Feb 2005
14. P. Sapaty, *Ruling Distributed Dynamic Worlds* (Wiley, New York, 2005)

Chapter 8
Parallel and Distributed Network Operations

8.1 Introduction

The chapter is devoted to solving some known theoretical graph and network problems where each graph node may be in a separate location, with links connecting them in the same or in different computing facilities.

Fully distributed and parallel solutions in SGL are shown for various paths between network nodes, which may take into account links orientation and certain constraints established on both links and nodes. Finding shortest path trees and shortest paths based on them are exhibited in SGL too, also shown is how the shortest path found can be used for optimal transportation of physical matter in a distributed space.

Solutions are offered for parallel finding of weak and strong graph components, like articulation points and cliques, which may have important applications in large networked systems influencing heir local and global safety and productivity. Of particular interest may be finding arbitrary network substructures in large distributed networks for which universal solutions in SGL are offered operating in fully distributed mode, without central resources. Finding some specific structures in arbitrary networks, which may be of practical use, is shown too.

The presented scenarios, being extremely compact, can effectively involve into operation and cooperation the whole network resources which may be in great numbers and spread worldwide. Solutions of other graph and network problems influenced by distributed processing methods and expressed in previous versions of SGL can be found elsewhere, especially in [1, 13].

Some scenarios were also inspired by parallel echo-based operations on general graphs [14], which appeared to be simpler and much more compact in our case due to shifting many traditional network-related operations to powerful automatic and distributed SGL interpretation, which had to be programmed explicitly otherwise. This is also true for comparison with other traditional methods of distributed solving of graph and network problems [15, 16].

© Springer International Publishing AG 2017
P.S. Sapaty, *Managing Distributed Dynamic Systems with Spatial Grasp Technology*, Studies in Computational Intelligence 690,
DOI 10.1007/978-3-319-50461-2_8

8.2   Finding Simple Paths Between Nodes

An example of a network on which we will be showing the finding of different paths is depicted in Fig. 8.1, where names of nodes and links can be any alphabetical and/or numerical strings, but for simplicity and compactness represented here by one or two characters only.

8.2.1   All Simple Paths

Beginning from a node which should be starting point of the path (let it be a), the following scenario self-spreads through the network in parallel and replicated manner passing nodes first time only and accumulating their names until reaches the node which should be the end of the path (let it be f), blocking further spreading from this final node (while other branches may continue finding other paths between a and f).

```
frontal(Path); #a;
output_repeat(
  notbelong(NAME, Path); append(NAME, Path);
  if(NAME == f, blind(Path)); all#)
```

Nine paths will be found between a and f and issued in node a:

```
(a, b, d, f), (a, b, d, e, f),
(a, b, d, c, e, f), (a, b, c, d, f),
(a, b, c, d, e, f), (a, b, c, e, f),
(a, c, b, d, f), (a, c, b, d, e, f),
(a, c, e, f)
```

Fig. 8.1 Network example for path finding

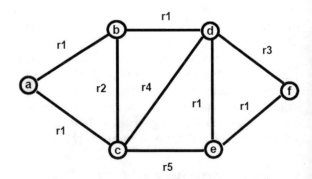

8.2.2 *Limited Length Simple Paths*

Let us restrict the allowed length of simple paths between given nodes, say, by 5, with the corresponding scenario following.

```
frontal(Path, Limit = 5); #a;
output_repeat(
 notbelong(NAME, Path); append(NAME, Path);
 if(NAME == f, blind(Path));
 count(Path) < Limit; all#)
```

The found paths will be as follows:

```
(a, b, d, f), (a, b, d, e, f), (a, b, c, d, f),
(a, b, c, e, f), (a, c, b, d, f), (a, c, e, f)
```

But with `Limit = 4` we will receive only:

```
(a, b, d, f), (a, c, e, f)
```

8.2.3 *Using Constraints on Links*

We can add any other constraints for finding paths, say, let them be of any length but only via links named `r1` or `r3`, as follows:

```
frontal(Path); #a;
output_repeat(
 notbelong(NAME, Path); append(NAME, Path);
 if(NAME == f, blind(Path)); (r1, r3)#))
```

For this case we will receive:

```
(a, b, d, f), (a, b, d, e, f)
```

8.2.4 Constraints on Both Links and Nodes

This case forbids passing links named r2 and r4 and also node named e, as follows:

```
frontal(Path); #a;
output_repeat(
  notbelong(NAME, Path); append(NAME, Path);
  if(NAME == f, blind(Path)); all#;
  notbelong(LINK, (r2, r4)); nonequal(NAME, e))
```

And as the result, we will receive only:

```
(a, b, d, f)
```

8.2.5 Taking into Account Orientation of Links

The same network but with both oriented and neutral links is depicted in Fig. 8.2.

Finding all paths in it from node a to node f along only oriented links will be as follows, where "along all" can be written as along_all, (along, all), or merely +all):

```
frontal(Path); #a;
output_repeat(
  notbelong(NAME, Path); append(NAME, Path);
  if(NAME == f, blind(Path)); + all#)
```

Fig. 8.2 Network with oriented and neutral links

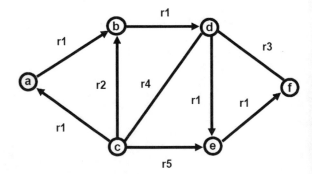

The result will be:

```
(a, b, d, e, f)
```

For a path against link orientation and also with non-oriented links, we will have:

```
frontal(Path); direct#a;
output_repeat(
  notbelong(NAME, Path); append(NAME, Path);
  if(NAME == f, blind(Path));
  (against, neutral, all)#)
```

With the obtained result: (a, c, d, f)

8.2.6   Issuing the Paths in the Final Node

The paths found from node a to node e in all cases above have been returned to node a (where the scenarios started) and issued there. But they can also be output in the final node f if this can be more suitable for the path finding application, as follows.

```
frontal(Path); direct#a;
repeat(
  notbelong(NAME, Path); append(NAME, Path);
  if(NAME == f, blind_output(Path)); all#)
```

8.3   Creating Shortest Path Tree

Finding Shortest Path Tree (SPT) covering all network nodes allows us to have a convenient structure containing shortest paths from the starting node to all other nodes. An example of SPT for the network with certain weights on links, starting with node a, is shown in Fig. 8.3 together with its orientation.

An SPT-finding scenario in SGL is as follows.

```
frontal(Far); nodal(Distance, Before);
#a; Distance = 0;
```

Fig. 8.3 Shortest path tree of
the network

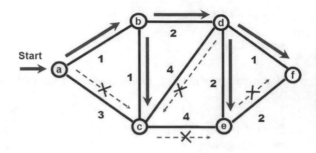

```
repeat(all#; Far += LINK
    or(Distance == nil, Distance > Far);
    Distance = Far; Before = PREVIOUS)
```

Starting in node a, it carries distance from it to other nodes (in frontal variable
Far) incrementing it with weights on the passed links (symbolizing distances
between neighboring nodes) and registering or updating it at nodes (in nodal
variables Distance), also each time fixing or updating the predecessor node this
distance came from (in nodal variable Before).

8.4   Finding Shortest Path Between Nodes

Having created the SPT starting in node a and covering all other nodes, we can
easily collect the Shortest Path (SP) from node a to any other node, let it be e in the
current case. The SP collection on SPT can start from a or e, with issuing the result
in any of these nodes. The solution, together with obtaining the SPT, can be single
source, starting in node a, or two source, starting in both a and e.

8.4.1   Single Source Solution

A single source solution for finding SPT and then SP between a and e is shown in
Fig. 8.4, which can be accomplished by the scenario that follows (using construct
quickest to abort unnecessary network activity after the path registration).

```
frontal(Far, Path); nodal(Distance, Before);
#a; Distance = 0;
sequence(
  repeat(all#; Far += LINK;
      or(Distance == nil, Distance > Far);
      Distance = Far; Before = PREVIOUS),
```

Fig. 8.4 Single-source finding of shortest path

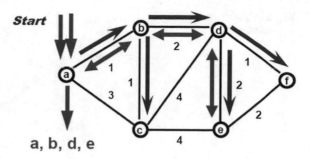

```
output_quickest repeat(
  Path & = NAME; if(NAME == e, blind(Path));
  all#; PREVIOUS == Before))
```

The result obtained in node a for the network of Fig. 8.4 will be:

```
a, b, d, e
```

The obtained path can also be issued in the final node e, as follows (aborting redundant network activity too, but using the state `fatal`):

```
frontal(Far, Path); nodal(Distance, Before);
direct#a; Distance = 0;
sequence(
 repeat(all#any; Far += LINK;
     or(Distance == nil, Distance > Far);
     Distance = Far; Before = PREVIOUS),
 repeat(
  Path & = NAME;
  if(NAME == e, (output(Path); fatal));
  all#; PREVIOUS == Before))
```

8.4.2 Two-Source Solution

This case supposes that SPT creation starts directly with node a, and SP collection starts directly from node e and collects SP by following it in the opposite direction, as in Fig. 8.5 and the following scenario, while outputting the collected SP in node a.

Fig. 8.5 Two-source finding
of shortest path

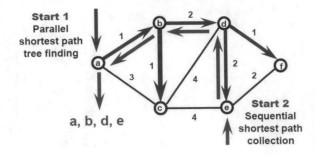

```
frontal(Far, Path); nodal(Distance, Before);
sequence(
(#a; Distance = 0;
repeat(all#; Far += LINK;
       or(Distance == nil, Distance > Far);
       Distance = Far; Before = PREVIOUS)),
(#e; repeat(Path = NAME & Path; #Before);
output(Path)))
```

The path collection, starting from the final node e and moving selectively up the
SPT and following a single-threaded path only, is much more economic as regards
the network activity than the previous case, which needs breadth-parallel propagation through the SPT unless the needed node is reached. But additional direct
access of node e may, from the other side, be more expensive than just having only
a single direct entry node a. (The spatial cycle collecting the path from node e
using rule `repeat` will automatically terminate in node a as there is no value in
nodal variable `Before`, as in the root of the tree.)

All solutions above have been found by navigating the network of weighed links
in parallel and distributed mode, without any central resources.

8.5   Moving Physical Matter via the Path Found

Let us consider moving physical matter (say, goods from production to consumption) via the shortest path found, assuming that nodes of the network are of
combined virtual-physical nature, i.e. having both names and physical coordinates.
We will access the current and previous node coordinates by environmental variables WHERE and LAST for obtaining real distances between them. Will also be
using special custom procedure `distance` (differing from the frontal variable
`Distance`) and `process` (the latter one related to the matter, assuming it needs
certain processing at the final location).

Fig. 8.6 Matter moving
along SP found

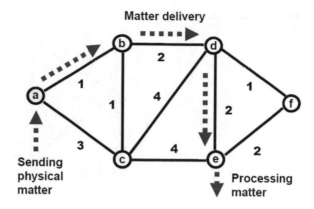

8.5.1   Matter Moving Along the Path

Let us consider moving the matter from node a to node e via the SP found, as in
Fig. 8.6 (assuming, for convenience, that weights at links reflect relative distances
between physical locations too, to have same SP picture as in Figs. 8.3, 8.4 and
8.5), using the following updated SPT scenario oriented on physical coordinates.

```
frontal(Far, Path); nodal(Distance, Before);
#a; Distance = 0;
sequence(
  repeat(all#; Far += distance(LAST, WHERE);
      or(Distance == nil, Distance > Far);
      Distance = Far; Before = PREVIOUS),
  Path = quickest_repeat(
    Path & = WHERE; if(NAME == e, blind(Path));
    all#; PREVIOUS == Before),
  (frontal(Cargo = "matter");
  repeat(move(withdraw(Path, 1))));
  process(Cargo)))
```

8.5.2   Matter Moving Opposite the Path

Another variant may represent an optimal movement of the matter in the opposite
way, i.e. from e to a, as shown in Fig. 8.7, directly via the shortest path between
them as part of SPT built from a. The following scenario will be doing this.

Fig. 8.7 Matter moving
opposite SP found

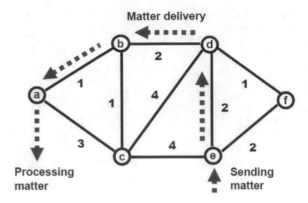

```
frontal(Far, Path); nodal(Distance, Before);
#a; Distance = 0;
sequence(
  repeat(all#; Far += distance(LAST, WHERE);
      or(Distance == nil, Distance > Far);
      Distance = Far; Before = PREVIOUS),
  (#e; frontal(Cargo = "matter");
  repeat_move(Before);
  process(Cargo)))
```

8.6   Finding Weak and Strong Components in Networks

In dealing with large distributed systems of particular interest may be finding their
weak and strong points and components, which may be used for solving different
optimization and security problems in them.

8.6.1   Finding Weakest Points

An example of weakest pointes in networked systems may be their *articulation
points* (one shown in Fig. 8.8), which when removed split the network into disjoint
parts.

The following scenario finds all such nodes in parallel and distributed mode,
resulting in an output of a single node name d for the current network (rule and
here operates on default as and_sequential).

Fig. 8.8 Network
articulation point

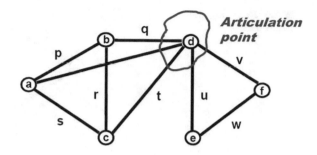

```
#all; COLOR = NAME;
and((random#; repeat(all#firstcome)),
    all#firstcome,
    output(NAME))
```

This distributed scenario works in the following steps:

- Starting in each network node, marking it with personal color, which may be any individual value, say, its NAME or ADDRESS (the latter may be preferable if the network node names are not unique, whereas addresses are). The environmental variable COLOR propagates and marks the visited nodes on the internal interpretation level.
- Parallel marking with this color all accessible network part via randomly chosen neighbor of the starting node, visiting the reached nodes only once and excluding the starting node from this navigation and marking process, as already visited and marked.
- Checking if the starting node solely connects parts of the network—this will take place if at least a single neighbor of the node is not marked with its personal color (i.e. has not been reached in the previous navigation process).

Having found the weak points in a distributed network, we can impact it, if needed, in different ways, depending on the application. For example, we can remove these points and weaken the network even more, as follows (what will remain of it after this is shown in Fig. 8.9).

```
#all; COLOR = NAME;
and((random#; repeat(all#firstcome)),
    all#firstcome,
    delete(NAME))
```

We can also try to strengthen the network by not allowing the articulation points found to split it when removed, say, by adding new links between other nodes. In the simplest case these can be among direct neighbors of the weakest nodes found,

Fig. 8.9 Removing the
weakest node

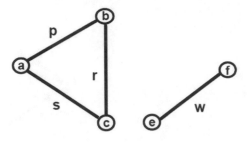

say, between the nodes first and last accessed from them in the previous scenarios
(a possible situation shown in Fig. 8.10 with new links named x added), as follows.

```
#all; COLOR = NAME; nodal(Next);
and((random#any; (#BACK; Next) = ADDRESS;
   repeat(all#firstcome)),
   (all#firstcome; linkup(x, (#BACK; Next)))))
```

8.6.2   Finding Strongest Parts

A strong, actually any, part of a networked system may mean a number of nodes in
it which are interconnected in certain way. In this respect *cliques* (or maximum
fully connected sub-graphs of a graph, as in Fig. 8.11) are often considered as the
strongest parts of a network.

They all can be found in parallel by the following simple scenario (establishing
lower threshold on the number of fully interconnected nodes as 3 to be considered
as a clique):

```
#all; COLOR = NAME; frontal(Clique = NAME);
repeat(all#firstcome;
```

Fig. 8.10 Strengthening the
network by additional links

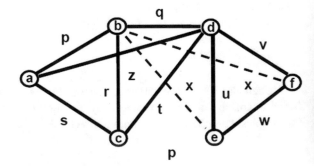

Fig. 8.11 Finding strongest parts, or cliques

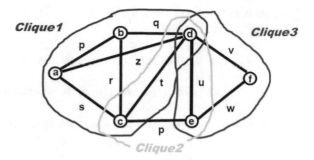

```
    if(and_parallel(any#Clique),
      if(BACK > NAME, Clique & = NAME, done),
      fail));
count(Clique) >= 3; output(Clique)
```

It will result in the following cliques for the network of Fig. 8.11:

```
(a, b, c, d), (c, d, e), (d, e, f)
```

The SGL scenario operates in the following steps:

- Starting in each node.
- Traversing links to neighbors and growing a potential clique in a unique node order (to block duplicates) until possible, where each new node added should have links with all previous nodes.
- Reporting the resultant clique with the threshold size given.

After finding the cliques, we can do different things for a certain impact on the network, say, removing some links between nodes in each of them. This will result in the network without cliques or with a fewer number of them and/or their reduced size, as in Fig. 8.12 and by the following scenario, where one link in each clique found is removed.

Fig. 8.12 Reduced and weakened cliques

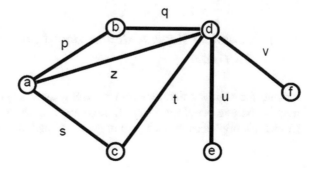

```
all#; COLOR = NAME; frontal(Clique = NAME);
repeat(all#firstcome;
    if(and_parallel(any#Clique),
      if(BACK > NAME, Clique & = NAME, done),
      fail));
count(Clique) >= 3; random#Clique; delete(LINK)
```

The modified network will still have two three-node cliques:

(a, b, d) ,   (a, c, d)

8.7   Finding Arbitrary Structures in Arbitrary Networks

We have considered finding parts of networks possessing certain qualities but which may have different number of nodes, structures, and locations. Let us consider now particular structures as the goal (represented as images or patterns) which should be found in arbitrary networks regardless of their sizes, topologies, and distribution in virtual and/or physical spaces.

8.7.1   Exemplary Network and Search Template

An exemplary network on which we will be showing such solutions is depicted in Fig. 8.13.

Any structures in distributed networked systems with any topologies can be found by describing them in SGL, like by the given template in Fig. 8.14. It can be applied from any network node and then, replicated in all nodes, self-evolves from them in parallel pattern-matching mode, finding meanings of variables X1–X6.

There may be a variety of possibilities of how to represent arbitrary search patterns/templates in SGL.

8.7.2   Template Representation Based on a Path Through All Nodes

Among the simplest methods (not always most time-efficient but fully universal) may be the one based on a path through all template nodes, as shown in Fig. 8.15 (for more complex cases such path may have to include some nodes more than once).

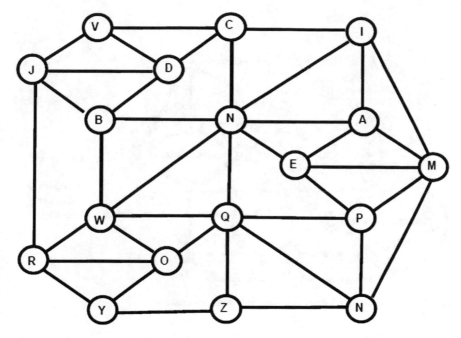

Fig. 8.13 Arbitrary network example

The SGL self-matching scenario based on the oriented path of Fig. 8.15 will be as follows.

```
#all; frontal(Match);
four(notbelong(NAME, Match); Match & = NAME; all#);
state(any#Match[1]); Match & = NAME; all#;
state_andparallel(any#Match[1,2,3]); Match & = NAME;
output(Match)
```

The nodes of three found substructures corresponding to variables X1 to X6 will be as follows:

```
(X1, X2, X3, X4, X5, X6)  →  (J, V, C, N, B, D),
                             (M, A, N, Q, P, E),
                             (R, W, Q, Z, Y, O)
```

The output of contents of variables found will be accomplished in nodes corresponding to variable X6, which will be nodes D, E, and O of Fig. 8.14.

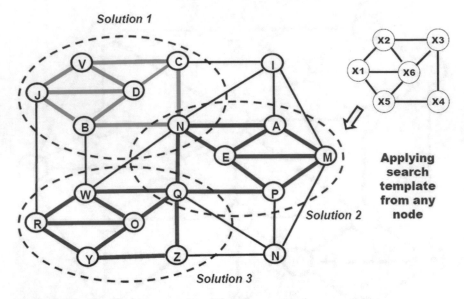

Fig. 8.14 Matching of arbitrary template with arbitrary network

Fig. 8.15 A path through all
nodes of the template

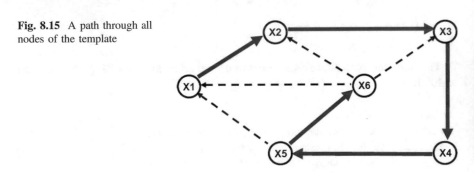

8.7.3 *Template Representation Based on a Path Through All Links*

Another simple template representation in SGL, fully universal too, is based on a
path through all template links, as in Fig. 8.16 (with numbers on links corresponding to the order of their passing), where in general some links may have to be
passed more than once. The corresponding SGL solution follows.

```
#all; frontal(Match = NAME);
five(all#; notbelong(NAME, Match); Match & = NAME);
#BACK; all#Match[1]; all#Match[6];
```

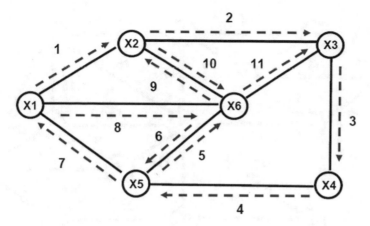

Fig. 8.16 A path through all links of the template

```
all#Match[2]; #BACK; all#Match[3];
output(Match)
```

The solution for the template variables found (same as above for the path via all nodes) will be issued in nodes corresponding to variable X3, which will be C, N, and Q of Fig. 8.14.

8.7.4   Networks with Named Nodes and Links

Let us consider now a network with both nodes and links having names, as in Fig. 8.17.

And of interest may be collecting jointly names of nodes and links in the matches found, having variables on both nodes (like X1 to X6) and links (like L1 to L9), as in Fig. 8.18.

The corresponding SGL scenario for finding and issuing values of all these variables in the successful matches will be as follows.

```
#all; frontal(Nodes = NAME, Links);
five(all#; notbelong(NAME, Nodes);
    Links & = LINK; Nodes & = NAME);
#BACK; all#Nodes[1]; Links & = LINK;
all#Nodes[6]; Links & = LINK;
all#Nodes[2]; Links & = LINK;
#BACK; all#Nodes[3]; Links & = LINK;
output(Nodes & ':' & Links)
```

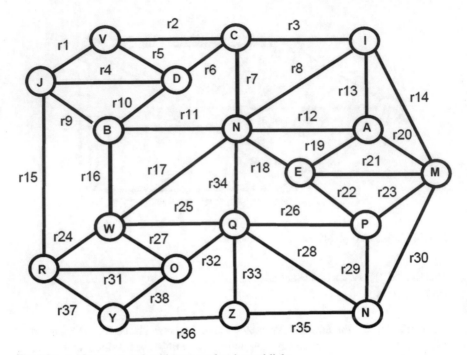

Fig. 8.17 Arbitrary network with names of nodes and links

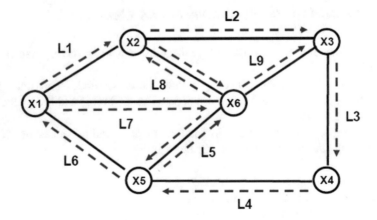

Fig. 8.18 Search template with variables on both nodes and links

The following solution for variables at both nodes and links will be issued in nodes corresponding to variable X3, which will be C, N, and Q of Fig. 8.17, same as in the previous example.

```
(X1, X2, X3, X4, X5, X6):
 (L1, L2, L3, L4, L5, L6, L7, L8, L9) →
  (J, V, C, N, B, D):
   (r1, r2, r7, r11, r10, r9, r4, r5, r6),
  (M, A, N, Q, P, E):
   (r20, r12, r34, r26, r22, r23, r21, r19, r18),
  (R, W, Q, Z, Y, O):
   (r24, r25, r33, r36, r38, r37, r31, r27, r32)
```

8.7.5 Working with Networks Having Multiple Links Between Nodes

Not to have multiple solutions (for templates with node variables only) in case of existence of multiple links between same nodes, we can use the `neighbors` option when navigating the network, which allows us to move to neighboring nodes only once, regardless of the number links leading to them, as follows for a template based on a path through all nodes:

```
#all; frontal(Match);
four(notbelong(NAME, Match); Match & = NAME;
   neighbors#);
state(any#Match[1]); Match & = NAME; neighbors#;
state_andparallel(any#Match[1,2,3]); Match & = NAME;
output(Match)
```

But for finding matches with both node and link names we should allow multiple solutions which may have different link names in them with the same nodes.

8.8 Examples of Finding Particular Structures

Let us consider a few examples of finding simple patterns in networks containing certain symmetry, and which may need to be issued as unique images without duplicates (for this, some additional measures should be included into the search patterns).

8.8.1   Example 1: Triangle

The simplest pattern of such type may be a triangle (shown in Fig. 8.19) which can be found starting from its any node while navigating other nodes in two different directions.

By establishing the following constraint on names (if unique in the network) or addresses of the nodes like:

```
X1 > X2 > X3
```

we will be able to get a unique match of its variables, as by the following scenario.

```
#all; frontal(Triangle = NAME);
2(all#; PREDECESSOR > NAME; Triangle & = NAME);
any#Triangle[1]; output(Triangle)
```

The following matching will be issued in the scenario starting nodes (for the network in Figs. 8.13, 8.14, and 8.17):

```
(X1, X2, X3) →
   (D, J, V), (B, D, J), (C, D, V),
   (C, I, N), (A, I, N), (A, E, N), (A, E, M),
   (E, M, P), (M, N, P), (N, P, Q), (N, Q, Z),
   (O, R, Y), (O, R, W), (O, Q, W), (N, Q, W),
   (B, N, W)
```

There may be more than a single triangle issued for the same nodes if the latter can be connected by more than one link, and such triangles will be link-based. We may use neighbors modifier in this case too, to protect from duplicates.

Fig. 8.19 Triangular pattern

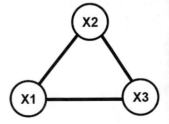

8.8.2   Example 2: Two Triangles Sharing a Side

Another, a bit more complex example is shown in Fig. 8.20 combining two triangles sharing a side (or being a four-node figure with a single diagonal):

It, too, can be started from different nodes and propagated in different directions,
but by setting such constraint as X1 > X3 and X2 > X4 we can get a unique match
by the following scenario (in which we can also use the effective internal
node-coloring mechanism):

```
#all; frontal(Pattern = NAME); COLOR = NAME;
2(neighbors#firstcome; Pattern & = NAME);
state(any#Pattern[1]); Pattern[1] > NAME;
neighbors#firstcome; Pattern[2] > NAME;
Pattern & = NAME; any#Pattern[1];
output(Pattern)
```

Result issued (in nodes corresponding to variable X1):

```
(D, B, J, V), (D, C, V, J), (O, W, R, Y),
(O, Q, W, R), (Q, N, W, O), (N, B, W, Q),
(E, A, M, P), (A, M, I, N), (I, A, N, C),
(M, E, P, N), (N, M, P, Q), (N, P, Q, Z)
```

Using construct neighbors from the beginning instead of links, which may be
multiple between nodes, we can completely avoid duplicates in the results found.

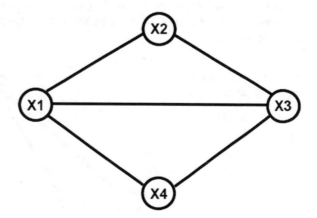

Fig. 8.20 Pattern with two
triangles sharing a side

8.8.3   Example 3: Unlimited Expanding Structure

One more example is shown in Fig. 8.21, which is to find maximum possible chains of nodes each of which also has a link with the same, let it be starting, node.

To dissociate from the two previous examples, we establish here the allowed lower number of nodes in the pattern as 5, and also set up constraint X2 > Xn allowing the chain to be passed only in one direction during its growth in the network, as follows.

```
#all; frontal(Pattern = NAME); nodal(Threshold = 5);
neighbors#; Pattern & = NAME;
repeat(neighbors#; notbelong(NAME, Pattern);
    state(any#Pattern[1]); Pattern & = NAME);
count(Pattern) >= Threshold;
Pattern[2] > Pattern[last]; output(Pattern)
```

The issued result for the network of Figs. 8.14 and 8.17 will be as follows:

```
(D, B, J, V, C), (N, C, I, A, E),
(M, I, A, E, P, N), (N, M, P, Q, Z),
(W, B, N, Q, O, R), (O, Q, W, R, Y)
```

Much more complex templates with symmetries, alternatives, fuzziness, repetitions, etc. which may appear to be of particular interest in distributed systems, can be readily described in SGL and found in parallel and distributed manner, without any central resources.

In the examples above for finding certain parts of the network we always started from all network nodes to grow all possible solutions in them. In the real world, with possible millions and billions of interconnected network nodes this may happen to be costly and not needed in general, as we could often outline particular network regions and even separate nodes or their types to start with for solving concrete problems.

Fig. 8.21 An unlimited expanding pattern

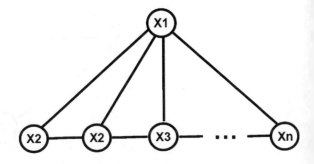

8.9 Conclusion

We have shown a variety of operations in SGL on distributed networks ranging from finding different paths in them to discovering parts with required qualities to grasping spatial images of arbitrary or specific structures. All such operations can work in parallel and fully distributed manner regardless of the size and structure of networks, also their distribution in virtual and physical spaces, which can be worldwide. As already mentioned, the SGT model represents operations on distributed systems in the form of active spatial patterns self-matching and grasping the system's body, and reflecting, in some sense, the natural brain's capabilities of perceiving visual images as a whole. That is why scenarios in SGL for solution of graph and network problems often appear to be as natural and simple as, for example, routine arithmetic or logic expressions in traditional algorithmic languages. The SGL solutions are also much clearer and shorter than the expression of existing graph and network algorithms in usual languages, being at the same time fully distributed, with each node potentially residing in a separate computer.

References

1. P.S. Sapaty, in *A Structural Approach to Solving Tasks of the Graph and Network Theory*. Proceedings of the 5th All-Union School on Parallel Programming and High-Performance Systems, Kiev, Naukova Dumka, 1982 (in Russian)
2. P.S. Sapaty, in *On Efficient Structural Implementation of Operations on Semantic Networks*. Proceedings of the USSR Academy of Sciences. Technical Cybernetics, No. 5, 1983 (in Russian)
3. P.S. Sapaty, in *Active Information Field as a Model for Structural Solving of Tasks on Graphs and Networks*. Proceedings of the USSR Academy of Sciences. Technical Cybernetics, No. 5, 1984 (in Russian)
4. P.S. Sapaty, A wave language for parallel processing of semantic networks. Comput. Artif. Intell. **5**(4) 1986
5. P.S. Sapaty, in *The Wave Approach to Distributed Processing of Graphs and Networks*. Proceedings of the International Working Conference Knowledge and Vision Processing Systems, Smolenice, November 1986
6. P.S. Sapaty, in *The WAVE-1: A New Ideology and Language of Distributed Processing on Graphs and Networks*", Computers and Artificial Intelligence, No.5, 1987
7. P.S. Sapaty, WAVE-1: a new ideology of parallel processing on graphs and networks. Future Gener. Comput. Syst. **4** (1988) (North-Holland)
8. P.M. Borst, M.J. Corbin, P.S. Sapaty, in *WAVE Processing of Networks and Distributed Simulation*. Proceedings of the HPDC-3 International Conference, San Francisco, Aug 94, IEEE (1994), pp. 61-69
9. P. Sapaty, *Ruling Distributed Dynamic Worlds* (Wiley, New York, 2005)
10. P. Sapaty, *Mobile Processing in Distributed and Open Environments* (Wiley, New York, 1999)
11. P.S. Sapaty, in *Solving Tasks on Semantic Networks and Graphs by Active Distributed Structures*. Proceedings of the 3rd International Conference Artificial Intelligence and Information-Control Systems of Robots, Smolenice (Elsevier Science Publishers B.V., North-Holland, 1984)

12. P.S. Sapaty, in *Organization of Advanced ISR Systems by High-Level Networking Technology*, MMC, no. 1 (2016)
13. P.S. Sapaty, in *Providing Over-operability of Advanced ISR Systems by a High-Level Networking Technology*. SMI's Airborne ISR, 26th–27th Oct 2015 (Holiday Inn Kensington Forum, London, United Kingdom, 2015)
14. E.J.H. Chang, in *Echo Algorithms: Depth Parallel Operations on General Graphs, IEEE Transactions on Software Engineering* (Vol. SE-8, Issue: 4, July 1982), pp. 391–401
15. W. Fan, in *Graph Pattern Matching Revised for Social Network Analysis*, ICDT 2012, 26–30 Mar 2012, Berlin, Germany
16. M. Raynal, *Distributed Algorithms for Message-Passing Systems* (Springer, Berlin, Heidelberg, 2013)

Chapter 9
Solving Societal Problems

9.1 Introduction

This chapter is for solving different problems related to human societies and social relations between people, it also introduces a new trend in the area called *Human Terrain* dealing with behavior of social groups whereas geographic terrain is about physical spaces.

Many problems in social systems, which are often counterintuitive in nature [1], can potentially be investigated and solved using *social networks* consisting of human nodes (like individuals or their groups, which may be hierarchical), and various relations between them. These networks may be huge, covering countries or even the world as a whole, and their understanding and management by a localized human single or even collective mind (and even equipped with existing information technologies) is not easy.

SGT, on the contrary, for solving societal problems can invoke the whole social network with associated local powers, activities, and information resources, which can solve any problems under self-spreading holistic scenarios in SGL. Some shown practical examples inherit more theoretical solutions of the previous chapter, and other ones consider finding individuals as potential leaders or through which most of communications between network nodes proceed. More cases deal with regular checking of a varying distance between topological centers of different distributed communities which can change in time, alarming if a certain threshold is reached.

Another scenario relates to the Ambient Assisted Living project, where mobile intelligence in SGL is checking and tracing movement of elderly people in urban environments and providing medical support if needed. More solutions are linked with organization of relief measures after natural or manmade disasters, including guided collective evacuation from a disaster zone.

© Springer International Publishing AG 2017
P.S. Sapaty, *Managing Distributed Dynamic Systems with Spatial Grasp Technology*, Studies in Computational Intelligence 690, DOI 10.1007/978-3-319-50461-2_9

SGL interpreters installed in available and especially widespread electronic devices (like laptops, video cameras, smart phones, smart watches, etc.) can effectively convert the whole social system into a universal spatial supercomputer capable of solving complex optimization and security problems on its own distributed body, and without any central resources.

9.2   Social Problems and Social Networks

9.2.1   Social Problems Examples

Social Problem [2, 3] can be viewed as an undesirable condition affecting a significant number of people which can be solved or eliminated through collective social action. Only few from the list of 122 in [3] are listed below:

Age discrimination	Alcoholism	Control
Child abuse	Child labour	Disaster relief
Domestic violence	Drinking and driving	Drug addiction
Eating disorders	Environmental pollution	Equal pay
Law enforcement	Gang violence	Genetically modified food
Global warming	Gun control	Gun rights
Hate crimes	Health care disparities	HIV/AIDS
Homelessness	Human trafficking	Hunger, mass murder
Minimum wage	Obesity	Organ and body donation
Outsourcing jobs	Pollution	Racism
Right to work	Sexual harassment	Smoking/tobacco use
Suicide	Terrorism	Unemployment
Violence in schools	Voting restrictions	Workplace violence

Some self-explaining examples of social problems are shown in Fig. 9.1.

9.2.2   Human Terrain Concept and Its Relation to Social Problems

Human Terrain (HT) [4–8] is based on analyzing complex problems in and between human societies through their anthropological, cultural, ethnic, political, and communicational dimensions, with subsequent finding effective mechanisms for resolving crises by non-military means.

Geographic Terrain, or merely *terrain*, is used to represent vertical and horizontal dimensions of land surface (see Fig. 9.2). This is usually expressed in terms

Fig. 9.1 Some examples of social problems

Fig. 9.2 Examples of geographic terrain and its representation

of elevation, slope and orientation features. Terrain affects surface water flow and distribution. Over a large area, it can affect weather and climate patterns.

Human terrain dealing with human population (symbolic examples in Fig. 9.3), its culture and interactions is a new and rapidly growing field of research, having originated, as a term, from military operations [8].

All conflicts in social systems are about people: their behaviours, attitudes, fears, social structures, family and ideological ties and narratives. Understanding the human dimensions of conflict is therefore a critical determinant in preventing conflict, shaping it and influencing the actors involved. It contributes to strategic

Fig. 9.3 Human terrain images

awareness, ability to plan and execute operations, helps to identify threats and opportunities.

Human terrain is usually defined as characterizing cultural, anthropological, and ethnographic information about the human population and interactions within a certain area.

9.2.3 Social Networks and Their Representation

Examples of social structures commonly visualized through social network analysis [9–11] include media networks, friendship and acquaintance networks, collaboration graphs, kinship, disease transmission, sexual relationships, etc. Social network analysis has emerged as a key technique in modern sociology. It has also gained a significant following in sociology, anthropology, biology, economics, geography, history, information science, organizational studies, political science, social psychology, development studies, sociolinguistics, and computer science.

In social network representations, human individuals are usually viewed as nodes, and links are expressing different relations between humans in both virtual and physical worlds (for the latter, say, as physical distances), as in Fig. 9.4.

Such networks may be nested as nodes can represent not only human individuals but their specific groups or collectives too, with links to other individual or group nodes, and this nesting may be multilevel. Nodes, except names and other attributes may have addresses or positions in both physical and virtual spaces.

Many social problems can be formulated and solved on such networks. The networks may be very large, distributed, and active, changing in space and time. Usually, they cannot be comprehended from a single point, even in principle, and their effective parallel and distributed processing, management, and simulation are very important to be dealt with properly.

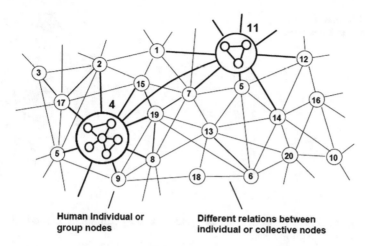

Fig. 9.4 Formal representation of social networks

9.3 Exemplary Social Network Operations

Some of the presented in this section examples in SGL inherit theoretical solutions on general graphs and networks considered in the previous chapter, being rather oriented here on practical tasks in social systems. More social networks-related scenarios in SGL can be found in [7] under the human terrain context.

9.3.1 Distributed Counting of the Number of Nodes and Links

An exemplary social (or human terrain) network on which we will be showing some solutions in SGL is shown in Fig. 9.5, with human nodes uniquely numbered

Fig. 9.5 Exemplary social network

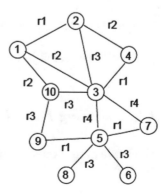

Fig. 9.6 The leader as the
node with maximum number
of links

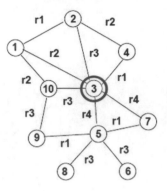

(named) from 1 to 10 and different relations between them named as r1–r4,
which can be repeating.

Counting the number of all nodes can be performed by the following elementary scenario starting in any node and operating in parallel and distributed way
by spatially navigating the network:

```
output(count(#all))
```

Result obtained in the starting node will be: 10.
Counting the number of all links may have the following expressions:

```
output(count(#all; all#) / 2 )   or
output(count(#all; all#; PREVIOUS > ADDRESS))   or
output(count(#all; all#; PREVIOUS < ADDRESS))
```

Result in the starting node will be: 14. The level of social system organization
can be roughly estimated as the number of its links divided by the number of nodes.

Finding possible leader as the node with maximum number of links to other
nodes (which may not always reflect the real situation, however), as in Fig. 9.6 and
by the scenario below.

```
output(max(#all; count(all#) & NAME) : 2)
```

The obtained result in the starting node will be 3.

9.3.2 Finding Paths Between Nodes

All simple paths between given nodes, letting the latter to be nodes 1 and 6 as in
Fig. 9.7.

Fig. 9.7 All paths between
nodes 1 and 6

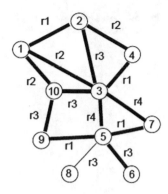

The solution of this task may hint on via which individuals you can potentially
reach one person from the other by using local relations between human nodes in
their network. The corresponding SGL scenario, starting in node 1, will be as
follows:

```
#1; frontal(Path);
output_repeat(
 notbelong(NAME, Path); append(Path, NAME);
 if(NAME = 6, blind(Path), all#))
```

All results will be output in the starting node 1:

```
(1, 3, 5, 6), (1, 2, 3, 5, 6), (1, 2, 4, 3, 5, 6),
(1, 3, 7, 5, 6), (1, 2, 3, 7 ,5, 6),
(1, 2, 4, 3, 7, 5, 6), (1, 10, 3, 5, 6),
(1, 10, 9, 5, 6), (1, 10, 3, 7, 5, 6),
(1, 3, 10, 9, 5, 6), (1, 2, 3, 10, 9, 5, 6),
(1, 2, 4, 3, 10, 9, 5, 6)
```

Finding all simple paths between given nodes via particular links. Let again
the start and final nodes be 1 and 6, and the allowed links between them as r2, r3,
r4 (see Fig. 9.8).

This details the previous task by explicitly naming links which should be passed
between nodes on the way to the target node.

```
#1; frontal(Path);
output_repeat(
 notbelong(NAME, Path); append(Path, NAME);
 if(NAME = 6, blind(Path), (r2, r3, r4)#))
```

Fig. 9.8 All paths between
given nodes via certain links

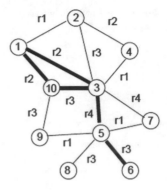

The result returned to the starting node 1 will be as:

```
(1, 3, 5, 6), (1, 10, 3, 5, 6)
```

**Finding all simple paths via particular links with collecting names of both
node and link passed** as an extension of the previous case, with SGL scenario
following.

```
#1; frontal(Path);
output_repeat(
  notbelong(NAME, Path); append(Path, NAME);
  if(NAME = 6, blind(Path));
  (r2, r3, r4)#; append(Path, LINK))
```

The result returned to the starting node 1 will be as:

```
(1, r2, 3, r4, 5, r3, 6),
(1, r2, 10, r3, 3, r4, 5, r3, 6)
```

Any single path between given nodes via any links with both node and link
names collected can be found as follows. A possible (among many) solution is
shown in Fig. 9.9 which is based on forming an implicit, hidden, spanning tree
covering the network during its parallel navigation, originating from the starting
node. The solution forbids repeated entry of the same nodes using special internal
interpretation mechanism triggered by modifier first. This may often provide the
quickest solution for a path between any two nodes, with only a single path produced by the following SGL code.

Fig. 9.9 Any single path
between nodes 1 and 6 via
any links

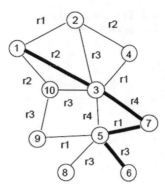

```
frontal(Path); #(first, 1);
output_repeat(
 append(Path, NAME);
 if(NAME = 6, blind(Path));
 all#first; append(Path, LINK))
```

A possible reply in the starting node 1:
(1, r2, 3, r4, 7, r1, 5, r3, 6) but not necessarily
(1, r2, 3, r4, 5, r3, 6) as the shortest path.

9.3.3   Shortest Path Tree and Solutions Based on It

Shortest Path Tree (SPT) covering all nodes. Such a tree starting from node 1 is
shown in Fig. 9.10.

The following scenario creates one of possible shortest path trees, starting from
the given node and covering all other nodes. In a networked social system this may
help a certain person to reach all other persons via existing interpersonal relations in
a quickest way.

```
#1; frontal(Length = 0); nodal(Distance);
repeat(
 or(Distance == nil, Distance > Length);
 Distance = Length; Length += 1;
 Previous = PREVIOUS; all#)
```

SPT will be recorded directly in the network structure by using nodal variables
Previous associated with each node and pointing at the predecessor nodes in the

Fig. 9.10 Possible shortest
path tree from node 1

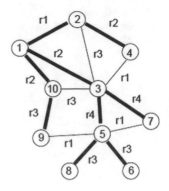

Fig. 9.11 Fixing shortest
path between nodes using the
SPT built

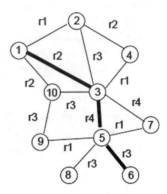

tree (the addresses of these predecessors were obtained via the environmental
variable PREVIOUS when staying in nodes).

Shortest Path (SP) between nodes based on SPT built. The created SPT
allows us to easily fix SP between its root in node 1 and any other node, as in
Fig. 9.11 between nodes 1 and 6.

Fixing shortest path between nodes 1 and 6 can be organized by using the
already created SPT from node 1 in a reverse order starting from node 6, or in a
straight order from node 1, with the result output in node 1 in both cases. If to start
in node 6, the full scenario including the SPT creation will be as follows:

```
#1; frontal(Length = 0); nodal(Distance);
stay_repeat(
  or(Distance == nil, Distance > Length);
  Distance = Length; Length += 1;
  Previous = PREVIOUS; all#);
#6; frontal(Path);
repeat(Path = NAME & Path; #Previous);
output(Path)
```

Reply in node 1 will be: 1,3,5,6.

Node from which SPT is lowest. This is a computationally intensive (despite working in parallel and fully distributed mode) task of creation of shortest path trees from every node and covering all other nodes, with subsequent choosing the node from which SPT appears having the lowest height. Such a node may have the highest probability to serve as *the leader of this group* of people. It can be found by the following SGL scenario.

```
output_min(
  #all; COLOR = NAME;
  frontal(Length = 0); nodal(Distance);
  stay_repeat(
    or(Distance == nil, Distance > Length);
    Distance = Length; Length += 1; all#);
  max(#all; Distance) & NAME)
```

The result will be (2, 3) with node 3 and SPT height from it as 2, being the lowest, see Fig. 9.12.

The longest among all shortest paths (also known as Diameter of the network) can be found by the following scenario:

```
output_max(
  #all; COLOR = NAME;
  frontal(Length = 0); nodal(Distance);
  stay_repeat(
    or(Distance == nil, Distance > Length);
    Distance = Length; Length += 1; all#);
  #all; Distance)
```

The obtained result will be 3.

Fig. 9.12 The node from which SPT has the lowest height

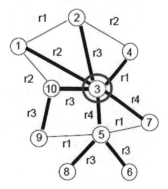

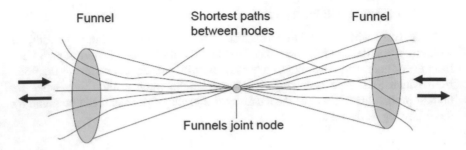

Funnel   Shortest paths   Funnel
                          between nodes

Funnels joint node

Fig. 9.13 Funneling and the funnels joint node

Funnelling relates to the situations where most information in social networks may happen to pass through a limited number of people, as in Fig. 9.13 for a single such node.

We may be interested in finding how many shortest paths between all possible couples of nodes in the network are going through a particular person (let it be node 10 in Fig. 9.14), and this can achieved by the following scenario with such a node represented as `Funnjoint`.

```
output_count(
   #all; COLOR = NAME; nodal(Before, Distance);
   frontal(Length = 0, Funnjoint = 10);
   stay_repeat(
    or(Distance == nil, Distance > Length);
    Distance = Length; Before = PREDECESSOR;
    Length += 1; all#);
   #all;
   repeat(if(NAME == Funnjoint, blind(1), #Before))
```

Result obtained in the starting node will be 15.

The strongest funnels joint node through which there exists maximum number of shortest paths between all other nodes can be found by the following scenario (finding such a node may be of particular importance when dealing with social networks, as it may represent their utmost strength or danger, depending on the applications):

```
output(
 max(
  #all; frontal(Funnjoint = NAME);
  count(
   #all; COLOR = NAME; nodal(Before, Distance);
   frontal(Length = 0);
   stay_repeat(
    or(Distance == nil, Distance > Length);
```

Fig. 9.14 Node 10 as the
funnel joint node

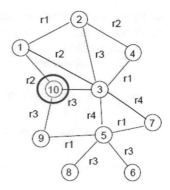

```
    Distance = Length; Before = PREDECESSOR;
    Length += 1; all#);
  #all;
  repeat(if(NAME == Funnjoint, blind(1), #Before)
  ) & NAME) : 2
```

The result issued in the starting node will be node 3.

9.3.4   Spatial Centres of Organizations

Finding spatial centre of a community. As mentioned before, the networked
human terrain model has both virtual and physical dimensions, where nodes may
have registered physical coordinates too. Using the latter, we may have general
impression about the physical area occupied by the social network or any other
generalized features, for example, the averaged topographical centre of the society
(as shown in Fig. 9.15).

The following SGL short scenarios can find such physical centre of the distributed social network without or with taking into account relations between nodes

Fig. 9.15 Averaged spatial
centre of the social network

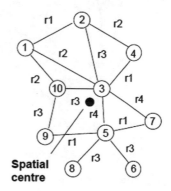

in the network. (Relations are increasing weight of nodes having more links with other nodes, and the integrated result combining virtual and physical weights of nodes may differ from just physical estimate.)

Links not accounted:

```
output_average(#all; WHERE)
```

Links taken into account (isolated nodes without links are dealt separately):

```
output_average(#all; or(all#, stay); WHERE)
```

Distance between spatial centres of organizations. This is based on the previous task of finding physical or combined physical-virtual centre of social network. We will first find centres of distributed (generally spatially overlapping) organizations by given types of links between their individuals (say, by ideologies, interests, or religions pursued), as in Fig. 9.16, where same individuals or groups may belong to different organizations simultaneously. We will then determine the physical distance between these centres of weight, which may hint, for example, on a possibility of certain tensions or even violence between these two communities, with the corresponding warning issued.

```
output_distance(
   average(#all; r1#; WHERE),
   average(#all; r2#; WHERE))
```

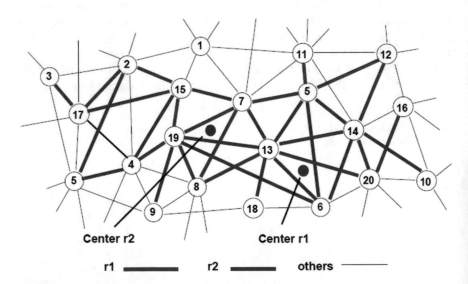

Fig. 9.16 Finding topological centres of different communities

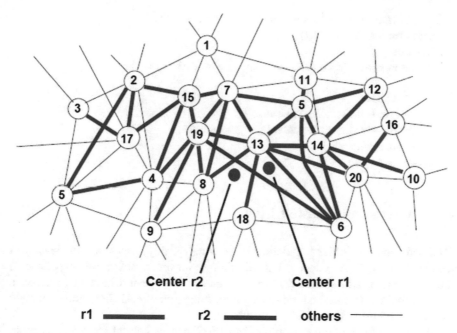

Fig. 9.17 Varying distance between different communities if nodes can move in time

Dynamic social networks with moving nodes. This is an extension of the previous task by regular checking the distance between the centres of two communities taking into account that nodes can move in physical space and the generalized distance between communities can change in time (see Fig. 9.17).

Regular checking of the averaged distance between the centres can be done by the following SGL scenario which also issues a warning when this distance falls below the Threshold given.

```
nodal(Threshold = 1 km);
sling(
 if(distance(
   average(#all; r1#; WHERE),
   average(#all; r2#; WHERE)) < Threshold,
  output('danger'));
 sleep(delay))
```

Regular distance checking in a distributed simulation mode. This combines distributed simulation of the social system allowing its networked nodes to randomly move in physical space while saving links with each other, with regular distance checking between the two centres of communities and issuing an alarm in case this distance becomes troublesome.

```
frontal(Range = (X(-50 m, 50 m), Y(-50 m, 50 m)),
nodal(Threshold = 1 km);
simulate(
 (#all; repeat(shift_random(Range); sleep(delay1))),
sling(
 if(distance(
    average(#all; r1#; WHERE),
    average(#all; r2#; WHERE)) < Threshold,
  output('danger'));
 sleep(delay2))
```

9.4 Active and Assisted Living

The percentage of older population is growing quickly, especially in developed countries [12]. The elders often remaining, however, the most experienced, knowledgeable, and skilful part of the society. European Union has created a special The Active and Assisted Living Joint Programme, AAL JP (former Ambient Assisted Living) [13] to tackle the problem.

We will show here an example of how SGT can be helpful for AAL JP, especially for overseeing and tracking individuals (say, elderly people) in large urban environments with the use of distributed sensor networks. The latter can be properly equipped and organized by spreading mobile intelligence in SGL, which can continually study the movement and behaviour of individuals and provide the necessary (like medical) support if needed.

Figure 9.18 shows some existing interface and visualization devices which when supplied with SGL interpreters can be used for AAL JP.

Tracking Individuals in SGL. The elderly people may happen to be lost, especially in crowded cities. The SGL scenario below shows how whereabouts of individuals can be regularly checked and traced by mobile spatial intelligence propagating in virtual world while following the movement of persons in physical world, as in Fig. 9.19.

Current positions of individuals can be lifted by their contacts with "semantic surfaces" [14] or by networked video cameras to which key pictures of the persons can be delivered by mobile intelligence accompanying them. The tracking intelligence can analyze and accumulate behavior of the moving person, demand checking her current physical condition (like heartbeat, blood pressure, body temperature, etc.). It can also alarm the nearest medical facilities in case of irregularities.

Many moving persons can be simultaneously and individually checked and served by the SGL scenario shown below, which can start in any network node

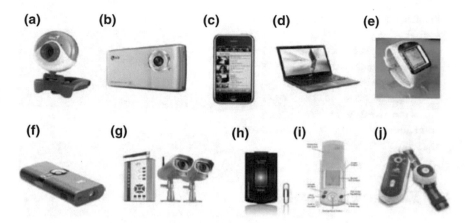

Fig. 9.18 Some tracking devices: **a** web camera **b** mobile phone camera, **c** mobile phone screen, **d** laptop keyboard and screen, **e** phone-watch camera, **f** portable projector, **g** wireless security cameras, **h–i** GPS-based tracers, **j** simple GPS-less gadgets for kids and elders control

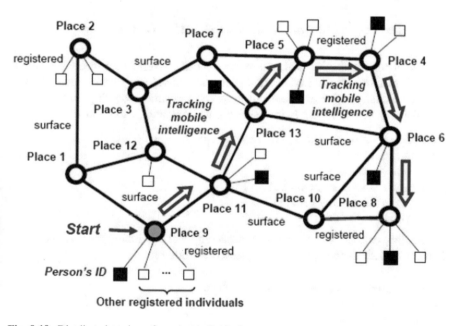

Fig. 9.19 Distributed tracing of moving individuals

where certain individual is supposed to be located at the current moment. The ID may contain the person's details (say, copy of identification documents and picture including), and History may be keeping selected medical data on this person, which can also be accessed in full via the internet address given.

```
frontal(ID = ..., History = ...);
repeat(
 loop(belong(ID, registered#);
    check_update(History, ID);
    if(problems(History),
      alarm_hospitalize(ID));
    sleep(delay));
 surface#; belong(ID, registered#));
alarm(ID & 'lost')
```

9.5   Emergency Management

We will consider here some scenarios related to emergency management, with more such tasks and their solutions (based on the previous versions of SGL) can be found in [15–19].

9.5.1   Investigating and Relieving Disaster Consequences

Embedded communicating SGL interpreters U can convert a post-disaster zone into a spatial machine capable of self-analysis and self-recovery under integral management scenarios. For example, all scattered casualties counting and returning scenario may be as follows (with its distributed operation shown in Fig. 9.20).

```
frontal(Area = disaster_area_definition);
output_sum(
 #(firstcome, Area);
 repeat(free_count(casualties),
     all#(firstcome, Area)))
```

Counting casualties in each region separately and organizing proportional relief delivery to each of them, may be expressed as follows:

```
frontal_Area = disaster_area_definition;
split_collect(
 #(firstcome, Area);
 repeat(free(count(casualties) & WHERE),
     all#(firstcome, Area)));
nodal(Region) = VALUE;
frontal(Supply) = multiply("relief", Region : 1);
move(Region : 2); distribute(Supply)
```

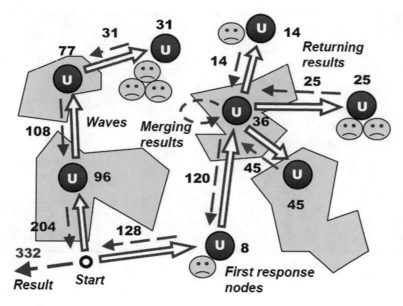

Fig. 9.20 Investigating and counting casualties

9.5.2   *Collective Evacuation from a Disaster Zone*

In case of major disasters (like earthquakes, hurricanes, flooding, landslides, forest fires, etc.) many people, especially elderly and handicapped, may need a special and urgent help to escape. A related SGL scenario setting up coordinated massive evacuation from the disaster zone may be activated by any person caught in such an event or by a special emergency organization.

This scenario can be regularly issuing instructions to individuals on where and how to move (say, via mobile phones if still working or any other channels). A chained collective movement through safe passage in a disaster zone may look like shown in Fig. 9.21.

Only the first individual in this chain is a pure leader (directly following the externally supplied waypoints) and the last one is a pure follower, whereas all others combine both functionalities (moving right after the previous person and directly followed by the next one).

```
cycle(
 exist(free_individual); N += 1;
  assign(free_individual, create_node(N)));
 (NAME == 1; Waypoints = (w1, w2, w3, ...);
  loop(output('move to:', withdraw(Waypoints, 1)));
```

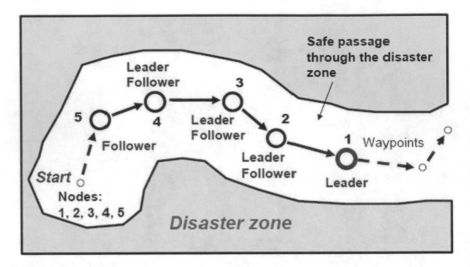

Fig. 9.21 Collective evacuation for a disaster zone

```
   wait(input == 'ok'))),
(NAME > 1;
sling(output('move to:', (#(NAME - 1); WHERE));
   wait(input == 'ok')))
```

9.6   Other Societal Tasks Currently Investigated Under SGT

We have shown above only simplest cases of using SGT and SGL for analyzing
and impacting social systems based on their networked representation. The SGL
scenarios presented can work in large distributed social spaces in highly parallel
and fully distributed mode, without central control, with communicated sensors
empowered by SGL interpreters embedded in human societies in large numbers
(even millions to billions). The following are some examples of other, more
complex, problems currently being investigated and programmed in SGL, which
may relate to solving societal problems on national and international levels.

- Investigating critical infrastructures and key resources of large societal systems
  in SGT with subsequent response to asymmetric situations;
- Providing global situational awareness in social systems by self-matching and
  self-recovering spatial intelligence in SGT;
- Investigation, analysis, and simulation of social networks dynamics;
- Investigating and blocking the spread of malicious ideas in human societies,
  especially those leading to massive violence and crime;

- Investigating ethnical, cultural, and religious origins of conflicts between groups of peoples and whole societies, which can reduce tensions and preventing conflicts between them;
- Using advanced unmanned systems for collecting smart data in human terrain systems which can simplify needed impact on them.

9.7   Conclusion

We have considered different kinds of tasks and problems emerging in modern human societies, which may be solved with the help of SGT. As was shown, many problems may be formulated and investigated by analyzing distributed social networks which can be processed in SGL in parallel and distributed mode if communicating copies of SGL interpreters installed in their key points, ideally in most of laptops, smart phones, smart watches, etc. Effective solutions have been shown for a number of network-formulated societal tasks, ambient assisted living supporting elderly people, and disasters and crises relief and management, all being compact and suitable for runtime generation and update.

References

1. J.W. Forrester, counterintuitive behavior of social systems. Technol. Forecast. Soc. Change **3**, 1–22 (1971–1972)
2. D. Michailakis, W. Schirmer, Social work and social problems: a contribution from systems theory and constructionism. Int. J. Soc. Welfare **23**(4), 431–442 (2014)
3. Social Problems and News Topics in the Twenty-First Century, Online Course Tutorial SOCY 105: Introduction to Contemporary Social Problems, University Libraries Guide. http://lib.guides.umd.edu/c.php?g=326995&p=2194601
4. R. Gonzalez, Human terrain: past, present and future applications. Anthropol. Today **24**(1), 21–26 (2008, February)
5. M. Montgomery, S. Fondacaro, Reflections on the human terrain system during the first four years. Prism **2**(4), 63–82 (2011, September)
6. K.E. Thuermer, Human terrain technology. Geospatial Intell. Forum (GIF) **11**(2). http://lib.guides.umd.edu/c.php?g=326995&p=2194601Thuermer
7. P.S. Sapaty, Distributed human terrain operations for solving national and international problems. Int. Relat. Diplomacy **2**(9) (2014, September)
8. Joint Doctrine Note 4/13, Culture and human terrain. Retrieved from https://www.gov.uk/government/publications/culture-and-human-terrain (2013, September)
9. R.S. Renfro, R.F. Deckro, A social network analysis of the Iranian Government. 69th MORS symposium, Working Group 8, 12–14 June 2001
10. W. Fan, *Graph Pattern Matching Revised for Social Network Analysis*. ICDT 2012, Berlin, Germany, 26–30 Mar 2012
11. C.T. Butts, Social network analysis: a methodological introduction. Asian J. Soc. Psychol. **11**, 13–41 (2008)

12. P.S. Sapaty, M. Sugisaka, *Advanced Networking and Robotics for Societal Engagement and Support of Elders*. Proceedings of 16th International Symposium on Artificial Life and Robotics (AROB 16th '11), B-Con Plaza, Beppu, Oita, Japan, 27–29 Jan 2011
13. The Active and Assisted Living Joint Programme (AAL JP), Digital Economy and Society. European Commission (2016). https://ec.europa.eu/digital-single-market/en/active-and-assisted-living-joint-programme-aal-jp
14. K. Dimitrov, K. Kanev, N. Mirenkov, Semantic surfaces in ambient assisted living environments. The 1st international workshop on aware computing, IWAC09, Aizu-Wakamatsu, Japan
15. P.S. Sapaty, M. Sugisaka, R. Finkelstein, J. Delgado-Frias, N. Mirenkov, *Emergent Societies: An Advanced it Support of Crisis Relief Missions*. Proceedings of Eleventh International Symposium on Artificial Life and Robotics (AROB 11th'06), Beppu, Japan, 23–26 Jan 2006. ISBN 4-9902880-0-9
16. P.S. Sapaty, R. Finkelstein, J. Filipe, *Distributed Emergency Management with Spatial Scenarios*. ed. by J.A. Cetto, J.L. Ferrier, J.D. Pereira, J. Filipe. Proceedings of Third International Conference on Informatics in Control, Automation and Robotics, ICINCO-2006, Volume: Intelligent Control Systems and Optimization, Setubal, Portugal, 1–5 Aug 2006, pp. 57–65. ISBN: 972-8865-59-7
17. P.S. Sapaty, M. Sugisaka, R. Finkelstein, J. Delgado-Frias, N. Mirenkov, Advanced IT support of crisis relief missions. J. Emerg. Manage. **4**(4), 29–36. ISSN 1543–5865 (2006, July/August)
18. P.S. Sapaty, Crisis management with distributed processing technology. Int. Trans. Syst. Sci. Appl. **1**(1), 81–92. ISSN 1751-1461 (2006)
19. P.S. Sapaty, M. Sugisaka, R. Finkelstein, J. Delgado-Frias, N. Mirenkov, Emergent societies: an advanced IT support of crisis relief missions. Artif. Life Robot. **11**(1). ISSN 1433–5298 (2007)

Chapter 10
Automated Command and Control

10.1 Introduction

The chapter investigates application of the developed approach to automated Command and Control (C2) for different distributed operations, with some previous publications related to this area in [1–4]. It starts with purely semantic scenario in a distributed environment with fully automatic C2 and overall system organization resulting from formal networked interpretation of SGL, which can be executed by different numbers of cooperating units of manned or unmanned origin. A detailed C2, if needed, can also be expressed explicitly and implemented in SGL, which is shown for distributed hierarchical and peripheral C2 infrastructures of a group of vessels in maritime environment.

Other described solutions in SGL relate to distributed tracing of multiple cruise missiles having complex and unpredictable routes but being kept in tight control by the spreading spatial intelligence individually tracing and following the missiles via organized sensor networks. The latter can be very useful in many other areas too, like the enhanced networked night vision of objects scattered throughout operational theatre. Also shown is how overall C2 of the planned European missile defense system can be automatically generated and maintained in SGL, allowing for individual tracing and supervision of multiple hostile objects on their full path from discovery to elimination. The chapter provides a comparison between the description of distributed operations in traditional battle management languages on the atomistic level of communicating military units with its holistic, semantic, equivalent in SGL, the latter being much simpler, shorter, and suitable for runtime modification.

And finally, it is demonstrated how SGT can be effectively applied not only to physically distributed systems but also to concentrated objects like aeroplanes or spaceships, where dealing with them in a fully distributed manner supported by automatic distributed C2 and without vulnerable central facilities can prove useful in critical situations.

© Springer International Publishing AG 2017
P.S. Sapaty, *Managing Distributed Dynamic Systems with Spatial
Grasp Technology*, Studies in Computational Intelligence 690,
DOI 10.1007/978-3-319-50461-2_10

10.2   Purely Semantic Scenario with Automatic Control

At the semantic level, we can describe in SGL what to do in a distributed space and the top decisions needed, regardless of a possible hardware or even entire system organization and management to accomplish this, which can be effectively shifted to automatic networked language interpretation with embedded formal and clear implementation rules.

10.2.1   Exemplary Task

A similar semantic-level task was first considered in relation to distributed sensors and cooperative robotics in SGL's predecessor DSL [5]. It is presented below in an extended variant and is formulated as follows.

Go to the following physical locations of the disaster zone with proximal coordinates:

50.433, 30.633
50.417, 30.490
50.467, 30.517

Evaluate damage in each location reached and get exact coordinates of the point where damage is measured, then find maximum among the obtained damage values and send the result together with the related exact coordinates to a management center.

The SGL expression for this task will be:

```
transmit_maximum(
  move((50.433, 30.633),
    (50.417, 30.490),
    (50.467, 30.517));
  assess(damage) & WHERE)
```

This compact semantic level task description includes both physical and virtual features, with potentially parallel physical forward movement in space, simultaneous local environment access, and computational echo operations. The obtained in different points and coupled together damage and coordinate values are returned and compared between themselves by their first part, which in case of maximum determines the resultant couple transmitted to the center.

This semantic description is fully defining of what to do in a distributed space and which decisions to make. It may be implemented with different numbers of doers which may be manned, unmanned or mixed. The solution with three doers is described below in detail where the overall system organization and management are fully in responsibility of communicating SGL interpreters installed in the doers, which are executing the SGL text by formal rules.

10.2.2   Three-Doer Task Solution

Imagine that three doers (as D1, D2, and D3) are present somewhere in the area containing the mentioned three locations, and the semantic scenario is initially injected from any doer like, for example, D1 as shown in Fig. 10.1.

A distributed group infrastructure is then created under the initiative of D1 which also preliminary invokes rules maximum and transmit, both being dormant at the beginning as their operands not ready yet, as shown in Fig. 10.2.

The scenario is also released from the mentioned outermost two rules embracing it, resulting in:

```
move((50.433, 30.633),
    (50.417, 30.490),
    (50.467, 30.517));
assess(damage) & WHERE
```

It is then converted into the following three independent branches by first splitting the compound move into three individual moves, by each coordinate, and then attaching to them the second, now becoming common grasp, as follows.

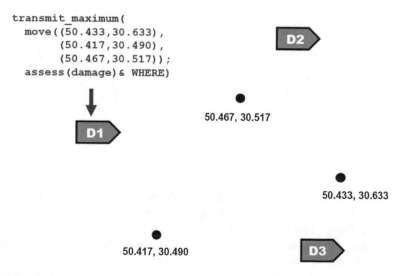

Fig. 10.1 Initial scenario injection into any doer

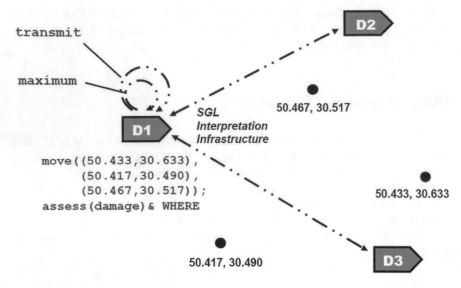

Fig. 10.2 Creating group infrastructure

Branch 1:

```
move(50.433, 30.633);
assess(damage)& WHERE
```

Branch 2:

```
move(50.417, 30.490);
assess(damage)& WHERE
```

Branch 3:

```
move(50.467, 30.517);
assess(damage)& WHERE
```

The obtained branches are then sent to the doers which appeared to be closest to the coordinates of rule move in the branches, as shown in Fig. 10.3, which begin interpreting them as individual scenarios.

Each doer is first interpreting the rule move causing independent physical movement to the selected point in space, carrying with it the second grasp to be used later, as depicted in Fig. 10.4.

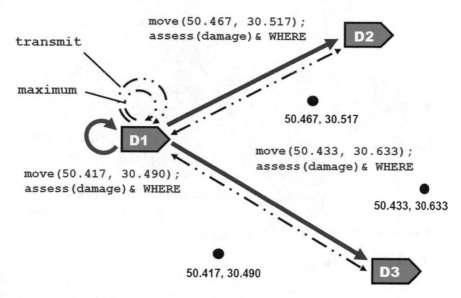

Fig. 10.3 Tasking all doers

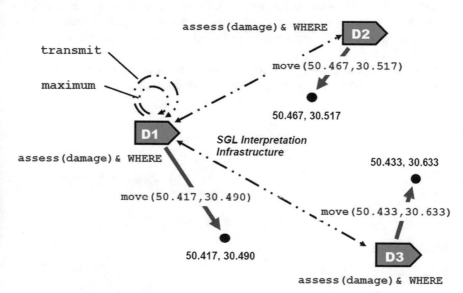

Fig. 10.4 Simultaneous doer movement

Having reached the dedicated destination points as close as possible, say, taking into account peculiarities of the terrain, each doer is then putting into operation the delivered second grasp (as in Fig. 10.5):

```
assess(damage) & WHERE
```

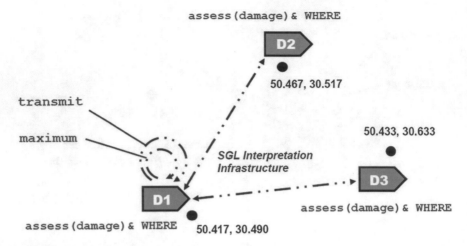

Fig. 10.5 Second step tasking

Each doer then assesses the damage (or hazardous conditions, as another variant) there using, for example, video cameras with image recognition techniques or measuring parameters like temperature or radiation, depending on the task details, as in Fig. 10.6.

Each doer also measures exact coordinates of the point reached using environmental variable WHERE, attaching them to the damage value obtained, as in Fig. 10.7.

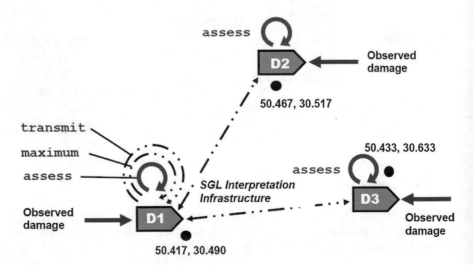

Fig. 10.6 Simultaneous damage assessment

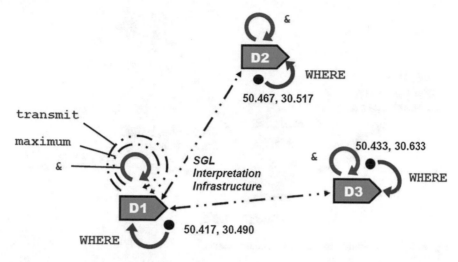

Fig. 10.7 Updating and attaching coordinates

The obtained two-value packets from the doers are sent via created group infrastructure to doer D1 (the latter already having its own result), and D1 finds maximum among the first values of the three packets received using initially invoked rule maximum, as in Fig. 10.8.

The packet with maximum observed damage value and related coordinates is sent by D1 using wireless link to a control centre responsible for this campaign, using previously engaged but remaining dormant rule transmit, as in Fig. 10.9.

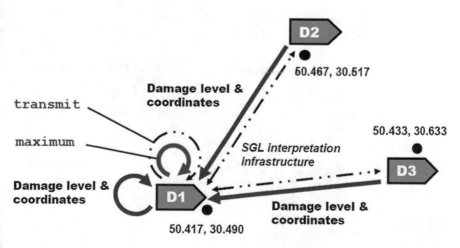

Fig. 10.8 Finding maximum damage value with attached coordinates

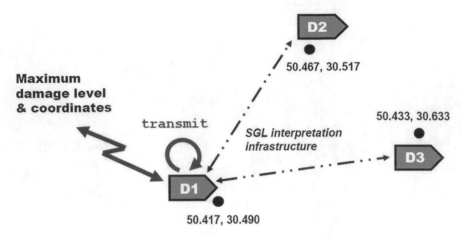

Fig. 10.9 Transmitting final result to a control center

10.2.3   Task Solution with Other Numbers of Doers

With other available numbers of doers having SGL interpreters installed, which can communicate with each other, the above semantic scenario, without any modification, can be executed cooperatively by them too, with quite a different jobs distribution, synchronization, and overall management, however, but everything can be automatically organized by the distributed interpreter's internal mechanisms.

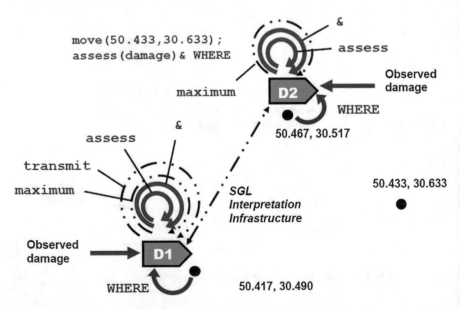

Fig. 10.10 Both doers are reaching the points initially selected and assessing damage there along with determining exact coordinates

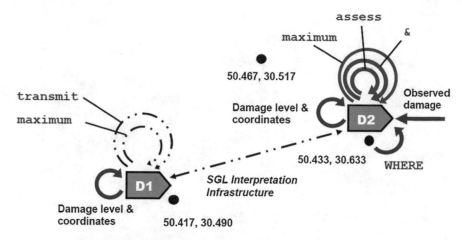

Fig. 10.11 The second doer moves to the third point (being closer to it than D1, and that is why it was chosen for this), measures damage and exact coordinates there. It also finds maximum between the damage values of the previous and current points, while the first doer is waiting for the receipt of this local maximum to compare with the damage value found by itself in its own world point, in order to get the global maximum

Figures 10.10 and 10.11 depict some intermediate snapshots of this task execution by only two communicating doers, where they have to perform more distributed loads, computational including, than in the described above case with three doers, in order to perform the task most efficiently.

10.3   Dynamic Creation of Distributed Command Infrastructures

In the previous section we considered purely semantic level of the task definition, with C2 and overall system organization handled by the distributed SGL interpreter fully automatically. But if needed, we can create in SGL and manage any infrastructures explicitly, as discussed in [6].

Imagine there is a fleet of sea vessels distributed over some area, and there also exist hostile objects that can be classified as targets to be eliminated, as shown in Fig. 10.12.

We will show the creation of two types of infrastructures which can help protect this fleet: a hierarchical one covering all nodes, and peripheral one, the latter for perimeter vessels only, say, to protect the whole group from outside.

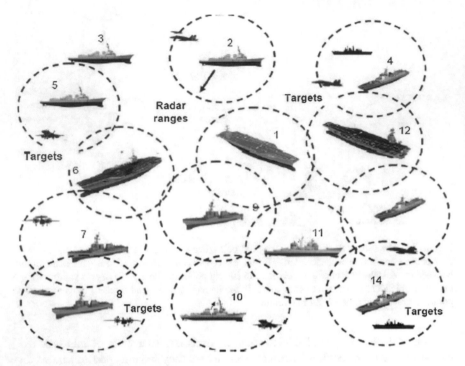

Fig. 10.12 Distributed fleet with hostile objects

10.3.1   Hierarchical Operational Infrastructure

An example of hierarchical infrastructure is shown in Fig. 10.13, starting from a provisional command centre unit 1 and covering all other units.

This distributed infrastructure creation and operation scenario in SGL may be as follows, which forms infrastructure links infra based on closeness of units to each other, with threshold *allowed_distance* for units to be considered as close.

```
#1; frontal(Seen, Depth = allowed_distance);
stay(repeat_linkup(+infra, Depth, firstcome));
sling(
  Seen = repeat(free_detect(targets), + infra#);
  repeat(free_select_impact(Seen), + infra#))
```

This self-evolving spatial scenario starts from the component selected as top of the hierarchy, as in Fig. 10.14.

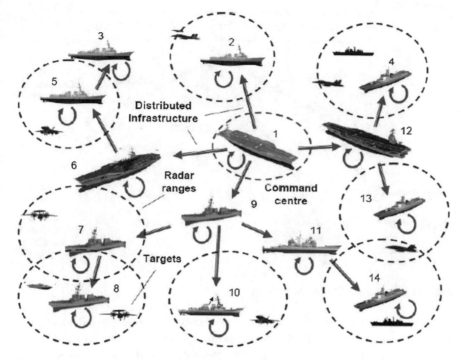

Fig. 10.13 Hierarchical operational infrastructure

Fig. 10.14 Initial scenario
injection

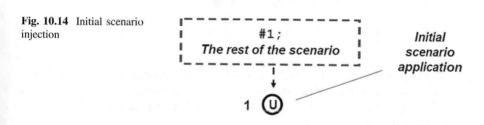

It creates persistent hierarchical infrastructure covering all nodes in a stepwise
top-down breadth-parallel manner by the following scenario fragment, and as in
Fig. 10.15.

```
repeat_linkup(+infra, Depth, firstcome)
```

Triggered by the internal SGL interpretation mechanisms with modifier
firstcome, the selected nodes can be visited only once thus preventing this parallel spatial process from cycling, always guaranteeing the tree-shaped resultant
infrastructure with top-down oriented links infra.

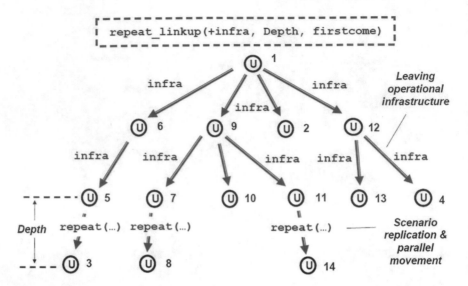

Fig. 10.15 Stepwise infrastructure creation

The scenario then uses the created infrastructure in top-down parallel navigation mode to activate all units and detect targets throughout the whole region, collect the targets and merge them in a parallel bottom-up echo mode via the infrastructure, and store altogether in frontal variable Seen at the headquarter node 1, as in Fig. 10.16 and by the scenario fragment below.

```
Seen = repeat(free_detect(targets), + infra#)
```

The targets collection by units is organized independently from the continuing global top-down units activation process via the infrastructure, with the use of additional rule free, but the discovered targets will be subsequently all merged by the bottom-up collection process.

The collected targets in frontal variable Seen at the headquarters node are then replicated and delivered to all units by parallel top-down spatial process using the created infrastructure where units, each having now full information about all targets, individually select the most convenient ones to shoot, as in the Fig. 10.17 and the scenario fragment that follows.

```
repeat(free_select_impact(Seen), + infra#))
```

The top-down targets delivery to all units is organized in parallel with possible targets impact operation by each unit, which is performed independently from the targets distribution process, as soon as it receives the targets from the latter, which is managed with the use of additional rule free.

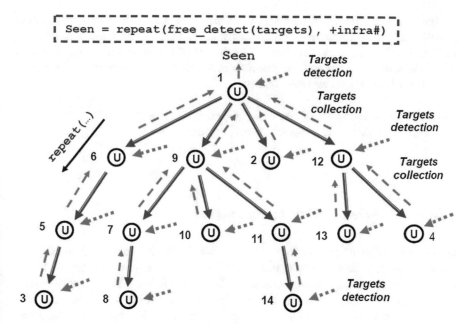

Fig. 10.16 Simultaneous infrastructure navigation, targets detection and collection

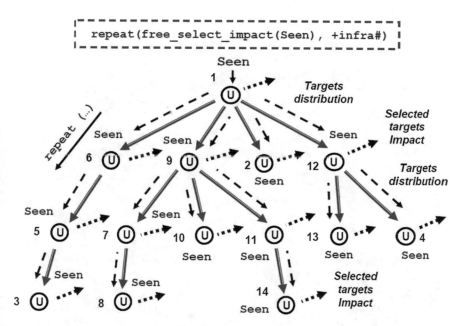

Fig. 10.17 Simultaneous targets distribution, selection and impact

The infrastructure-based parallel operations of spatial collection of distributed targets and their subsequent spatial distribution back to individual units with independent impact, shown in Figs. 10.16 and 10.17 can be organized together in a repeated mode using the most appropriate here rule sling, as in the full scenario text shown before, also repeated below:

```
sling(Seen = repeat(free_detect(targets), + infra#);
    repeat(free_select_impact(Seen), + infra#))
```

10.3.2   Peripheral, Ring Infrastructure

An example of peripheral, or ring, infrastructure (dedicated, say, for defending the whole distributed fleet against external, outside targets) is shown in Fig. 10.18. Its creation and operation can be accomplished by the following SGL scenario where the units to be included into this infrastructure are named explicitly (they can also be found automatically in SGL by analysing the units distribution in space and choosing outermost ones).

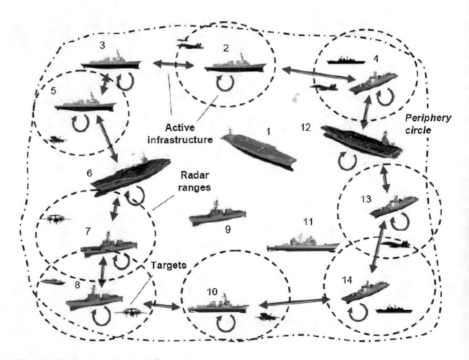

Fig. 10.18 Peripheral, ring infrastructure

```
frontal(
 Periphery = (4,12,13,14,10,8,7,6,5,3,2), Transit);
nodal(Seen);
#2;
repeat(
 free_loop(
 Seen = detect_merge(targets);
 select_impact(Seen),
 (Transit = Seen; infra#; merge(Seen, Transit))),
 linkup(infra, withdraw(Periphery, 1)))
```

This self-evolving spatial program, starting from any periphery unit (chosen as unit 2 in our case), creates ring infrastructure covering the peripheral units named by frontal variable Periphery and operates without any central control. The stepwise growing ring infrastructure without waiting for full completion (such overlap provided by the rule free) begins regularly collecting and distributing targets (to the neighbors) locally seen by individual units, leaving the final choice for attacks on the most appropriate targets to individual units, as in Fig. 10.19.

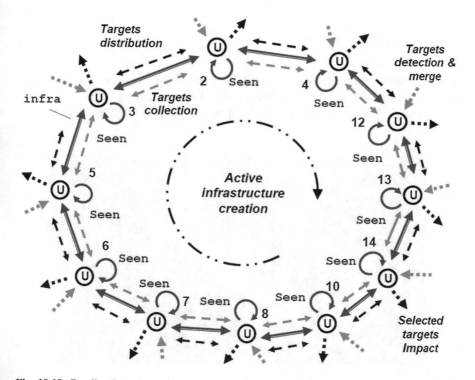

Fig. 10.19 Details of the ring infrastructure creation and operation

The targets seen and accumulated by individual units are being regularly exchanged with direct neighbors on the ring via the created links infra and merged with the targets they have discovered and accumulated themselves. This parallel and distributed process provides the flow of all targets via the entire ring infrastructure, thus supporting global awareness in each individual unit and by the system as a whole.

10.4   Withstanding Cruise Missiles

SGT can effectively organize discovery, tracing, analysing and proper impact of multiple low flying objects like cruise missiles, see Fig. 10.20, with unpredictable routes [7] by cheap distributed sensor networks operating under mobile spatial intelligence provided by the technology.

Cruise missiles have several advantages over ballistic missiles: they can be updated during flight, often pursuing complex routes to avoid detection. Their low flight altitude makes them very stealthy against air defence radars, and fuel efficient turbofan engines allow cruise missiles to be lighter and cheaper than their ballistic counterparts.

10.4.1   Existing Solutions

There are few but far from universal solutions for dealing with these types threats. Aerial sensors [8] are the best defence against low-flying cruise missiles, because they offer far better detection and tracking range than ground-based systems. The

Fig. 10.20 Tomahawk as a typical subsonic land-attack cruise missile

Fig. 10.21 Existing cruise missile defences: JLENS

bad news is that keeping planes in the air all the time is very expensive, and so are the aircraft themselves. Another solution is called Mountain Top [9], where high elevation points on the ground can be used to trace and target low flying missiles.

The primary challenge becomes the development of a reliable, affordable, long-flying, look-down platform. One that can detect, track and identify incoming missiles, then support over-the-horizon engagements in a timely manner. The Joint Land Attack Cruise Missile Defence Elevated Netted Sensor (JLENS) [10] is an example of such systems, as in Fig. 10.21.

The unmanned, tethered platforms can complement each other through the operation of both broad-area and precision radar systems, providing an over-the-horizon early warning capability.

10.4.2 Installing SGL Interpreters in Distributed Sensors

Embedding SGL interpreters into networked radar stations can convert the latter into universal distributed self-organized supercomputers capable of solving any problems within the space covered. These may include discovery, tracing, analysing, and destroying multiple aerial objects and low flying cruise missiles. Communicating radars can be effectively integrated with SGL interpreters in large environments of different natures and their combinations, like open land terrain, sea/ocean surface vessels, or urban environment.

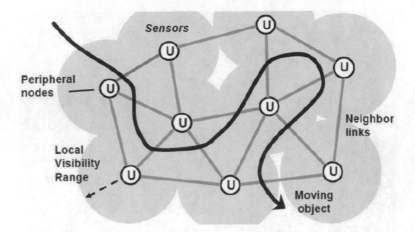

Fig. 10.22 SGL interpreters installed in communicating sensors with neighborhood links

Individual sensors have limited visibility range, but well organized distributed sensor networks empowered with SGT can provide continuous global vision of complexly moving objects through the space covered (as in Fig. 10.22), with their detailed study and destruction when required.

10.4.3   Distributed Missile Tracking Scenario in SGL

The SGL spatial tracking scenario may be as follows. Constantly operating in all region's peripheral sensors it catches an incoming object it sees and then follows wherever it goes with the help of individual mobile intelligence if the object is not seen from the current point any more (its visibility becomes lower than the given Threshold).

```
frontal(Object, Threshold = min_visibility);
hop(periphery, all);
whirl(
 nonempty(Object = search(aerial, Threshold, new));
 release_repeat(
  loop(visibility(Object) >= Threshold);
  max_destination(
   hop(neighbors, all); visibility(Object))))
```

Some stages of this distributed object tracking dynamics are shown in Figs. 10.23, 10.24 and 10.25, where spatial intelligence accompanying the

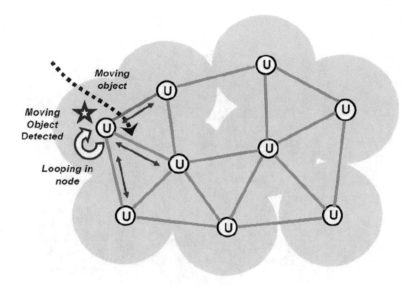

Fig. 10.23 Catching an alien object by a periphery sensor

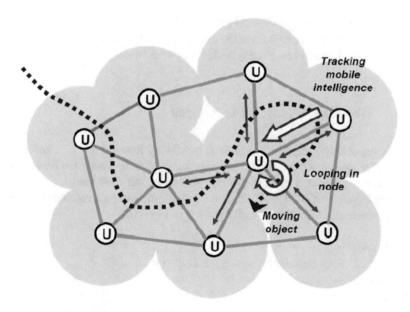

Fig. 10.24 Mobile intelligence keeping and following the moving object

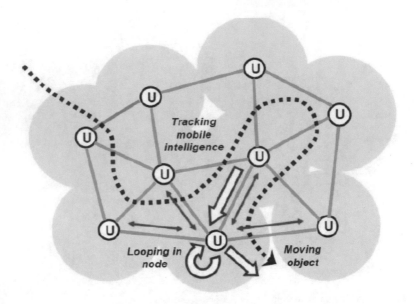

Fig. 10.25 Mobile object leaving the region if not destroyed

physically moving object via virtual networked space investigates the surrounding region if the object disappears from the current radar station, and then moves to the SGT-empowered neighboring radar where it is seen best.

10.4.4 Withstanding Multiple Attacks

Multiple, especially mutually coordinated attacks by low flying cruise missiles are considered at present as one of the most dangerous threats. These can be effectively handled (not only traced as in the previous section, but destroyed too) by mobile spatial intelligence of SGT with the use of distributed impact resources, as in the following scenario, also symbolically shown in Fig. 10.26.

```
nodal(Seen);
frontal(Object, Threshold = min_visibility);
hop(periphery, all);
whirl(
  Object = search(aerial, not_belong(Seen));
  visibility(Object) > Threshold;
  release(
   repeat(
    append(Seen, Object);
    loop(
```

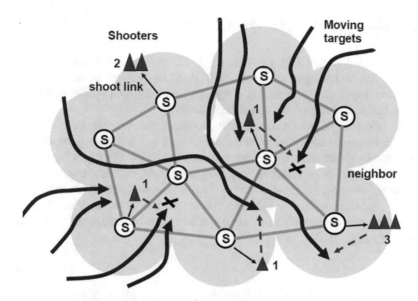

Fig. 10.26 Simultaneous tracing and destruction of multiple mobile targets in SGT

```
visibility(Object) > Threshold;
if((hop(shoot_link); CONTENT > 0;
   shoot(Object); decrement (CONTENT)),
   (withdraw(Object, Seen); done)));
withdraw(Object, Seen);
max_destination(
 hop(neighbor, all); visibility(Object));
if(visibility(Object) < Threshold),
   (output(Object & ' lost'); stop)))))
```

Each peripheral sensor is regularly searching for new targets, and each new target is assigned individual tracking intelligence which will propagate in distributed virtual space following the target's movement in physical space. If there are available shooters in the vicinity at each stage, a kill vehicle is launched against the target (with their available number is reduced afterwards). If the target is hit, it is removed form the further observation.

The scenario above can be easily extended for the case when different mobile intelligent branches evolving in space can cooperate with each other and with some global optimization processes, also in SGL, say, to optimize the use of limited impact resources scattered throughout the region of control of to identify and withstand multiple targets as the group ones with collective behavior.

10.5   Networked Night Vision Scenarios

Some examples related to night vision [11, 12] will be considered. Under SGT, multiple communicating sensors with SGL interpreters installed can be distributed throughout any regions, operating altogether as an integral spatial brain that can provide holistic vision of the operational theatre.

Such brain can help with solving ISR problems, for example, in case of reduced vision like at night or in bad weather, compensating insufficiency or distortion of sensor data by enhanced intelligence of the whole system. For example, allowing us to have a clearer (say, repeated and integrated from different angles) picture of separate objects or the theatre as a whole [12]. Different night vision equipment and scenario snapshots related to the tasks considered are shown in Fig. 10.27.

An example of scattered military objects and sensors throughout some region is depicted in Fig. 10.28.

10.5.1   Multiple Spatial Vision of a Particular Object

In Fig. 10.29 a case is shown where information of an object (symbolically named as R1 and given by exact coordinates) is being collected by all sensors covering it by their vision ranges and actually seeing, thus allowing us to obtain its multiple image from different angles.

Fig. 10.27 Night vision snapshots

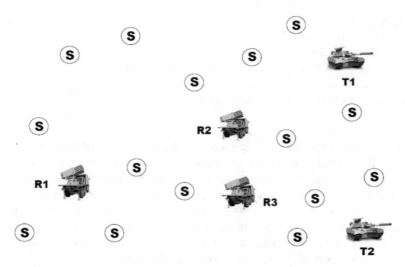

Fig. 10.28 Exemplary theatre with scattered sensors and military units

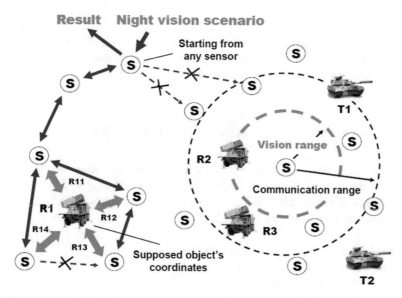

Fig. 10.29 Multiple vision of a single object

The corresponding SGL scenario may be as follows.

```
fontal(R1_coord = ..., Vision_range = ...,
    Communication_range = ...,
    Remote = infinite);
nodal(Remote1);
```

```
output_merge_repeat(
 Remote1 = distance(R1_coord, WHERE);
 Remote1 < Remote; Remote = Remote1;
 if((Remote < Vision_range; nonempty(Vision_range)),
   free_obtain(image, R1_coord));
 hop(all, firstcome, Communication_range));
```

The scenario starts from any sensor and propagates in parallel through the sensor network dynamically formed by the sensor communication ranges, to the region where local sensors, if exist, can potentially see the object within their vision ranges. It then activates the imaging rule obtain(image, R1_coordinates) in all such sensors if they actually see something by the coordinates given. The resultant image of R1 returned in multiple and differing from each other copies from the four prospective sensors of Fig. 10.29 will possibly merge into a single picture with the use of special technologies, as follows:

```
merge(R11, R12, R13, R14)
```

This final result will be returned to the sensor where the SGL scenario started. The resultant integral multi-view picture may happen to be even more useful than if received from a single sensor in normal vision conditions.

10.5.2   Multiple Spatial Vision of the Whole Theatre

To see the whole theatre equipped with night vision sensors, with possible multiple images of the same objects seen from different sides and angles if located within vision ranges of more than a single sensor, we may use this simple SGL scenario below with its work depicted in Fig. 10.30.

```
fontal(Vision_range = ..., Communication_range = ...);
output_collect_merge_repeat(
  if(nonempty(Vision_range),
    free_obtain(images : coordinates));
  hop(all, firstcome, Communication_range))
```

The resultant integrated images of the whole theater returned to the starting sensor will be as follows:

```
merge(R11, R12, R13, R14) : Coord_R1,
merge(R21, R22) : Coord_R2,
```

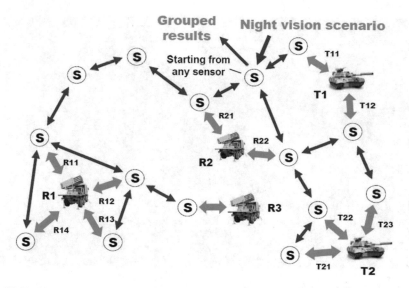

Fig. 10.30 Multiple vision of all objects

```
R3 : Coord_R3,
merge(T11, T12) : Coord_T1,
merge(T21, T22, T23) : Coord_T2
```

The scenario above, covering in parallel the whole sensor network by `Communication_range` of its sensors (while blocking possible cycling by the `firstcome` option in hops) collects the seen `images` within `Vision_range` of sensors together with their exact `coordinates`. This subsequently allows collected images with the same coordinates to be grouped together in the spatial echo `collect_merge` process, and finally output from the sensor where the scenario started.

10.6 Europe-Related Missile Defense Scenario

Let us consider here some scenarios relevant to the already discussed concept of a possible European missile defense system [13–19].

10.6.1 Missile Defense Main Stages

The missile defense system is supposed to work in the following stages.
Stage 1, shown in Fig. 10.31, with different steps numbered and having the following meaning.

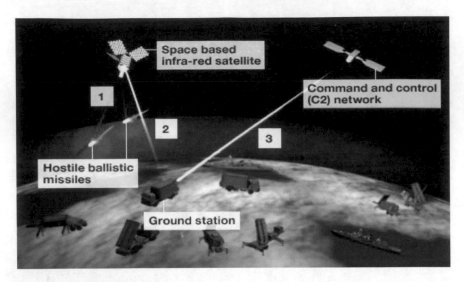

Fig. 10.31 Picking up a hostile missile

1. Infrared satellite system picks up heat signatures of hostile missiles launched towards target.
2. Information transmitted to ground stations for processing.
3. Processed information sent to C2 network.

Stage 2, where the C2 network relays information to sensor and weapons systems available in the region, as shown in Fig. 10.32.

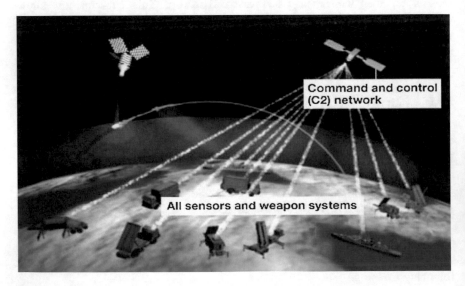

Fig. 10.32 Relaying the information obtained

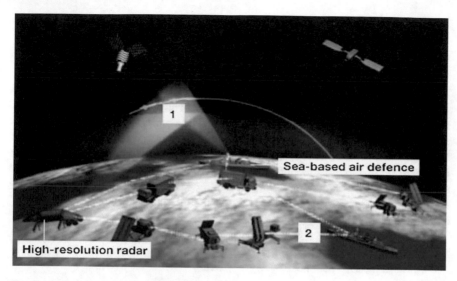

Fig. 10.33 Missile tracking & information sharing

Stage 3 develops in the following steps, depicted in Fig. 10.33.

1. Long-range sensors continue to track the missile to help command system calculate options for destroying them.
2. Information is constantly shared among the sensors and weapons systems.

In **Stage 4**, the command system has the option of shooting down the hostile missiles while in the upper or lower layers of the atmosphere, using corresponding upper or lower-layer available shooters, as in Fig. 10.34.

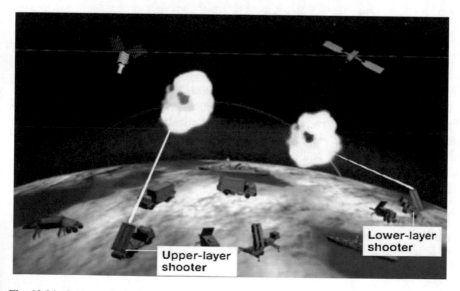

Fig. 10.34 Options of missile shooting

10.6.2 Missile Defense Management in SGL

We can symbolically extend the functionality and operations of the missile defense system mentioned above with possible Direct Energy Weapons, DEW (like high power lasers) located in space or on airborne (manned or UAV) platforms [19, 20].

Having synchronized DEW with infrared satellite sensors, we can write the whole SGL scenario integrating infrared satellites, DEW facilities, long range sensors, and upper and lower layer shooters into a dynamic flexible, distributed, and self-organized system. This system would be capable of discovering many hostile objects in parallel, simultaneously and individually tracing them at different stages of their flight, also launching (or re-launching in case of failures) proper impact facilities with verification of their success or failure until the targets are destroyed, as follows.

```
hop(all, infrared_satellite_sensors);
frontal(Target, Threshold = upper_lower);
sling(
 split_discover(missiles, new); Target = VALUE;
 release(
  sling(
   visible(Target); update(Target);
    if((hop(DEW); verify_shoot(Target)), done));
  hop(long_range_sensors);
  sling(
   visible(Target); update(Target);
   if(distsance(Target) > Threshold,
     hop(upper_layer_shooters),
     hop(lower_layer_shooters));
   if(verify_shoot(Target), done));
  output('Alarm! Target: ', Target, ' lost');
```

The advantages of this distributed management scenario are that it can be initially applied from any available system component, automatically creating distributed command and control infrastructure particularly oriented on the currently discovered targets and emerging situations. This automatically created distributed system organization can self-recover at runtime after indiscriminate damages to any system components (due to fully interpreted, mobile, virus-like implementation of SGL in distributed networked spaces).

In a broader scale, SGT can effectively integrate numerous distributed missile defence facilities and systems into global-goal-driven complexes operating under unified command and control, which can be fully automatic, especially in highly dynamic and asymmetric situations, as in Fig. 10.35.

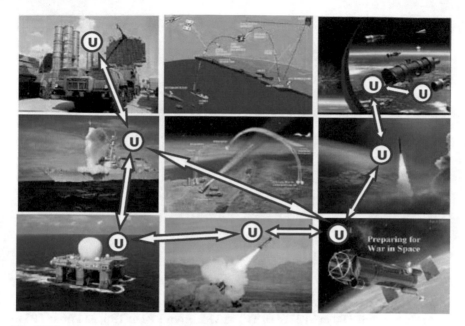

Fig. 10.35 Integrated global missile defense under SGT

10.7 High-Level Battle Management in SGL

Formalization of Command Intent (CI) and Command and Control (C2) in general are among the most urgent and challenging problems on the way to creation of effective multinational forces, integration of simulations with live control, and natural transition to robotized armies. Specialized Battle Management Languages for unambiguous expression of CI and C2 (like BML and its derivatives C-BML, JBML, geoBML, etc.) [21–24] are not programming languages themselves, needing therefore integration with other linguistic facilities and organizational levels to provide required system parameters.

On the contrary, working directly with both physical and virtual worlds, SGL allows for effective and universal expression of any battlefield scenarios and orders in parallel and fully distributed manner within the same, universal, language syntax and semantics; it also directly supports robotized up to fully robotic systems [1, 25]. And scenarios in SGL are much shorter and simpler than in BML-based languages.

10.7.1 Traditional Battle Management in BML

Let us consider an example taken from [22] and simplified in Fig. 10.36. The task is to be performed by two armoured squadrons BN-661 Coy1, and BN-661 Coy3,

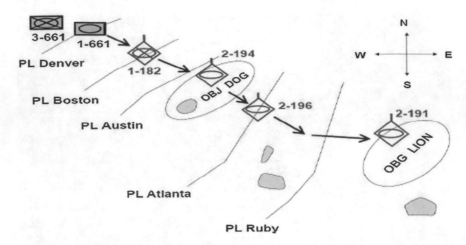

Fig. 10.36 An exemplary military scenario

which are ordered to cooperate in coordination. The operation is divided into four time phases: from TP0 to TP1, from TP1 to TP2, from TP2 to TP3, and from TP3 to TP4, to finally secure objective Lion, and on the way to it, objective Dog. Their coordinated advancement should be achieved by passing Denver, Boston, Austin, Atlanta, and Ruby lines, while fixing and destroying enemy units Red-1-182, Red-2-194, Red-2-196, and Red-2-191.

Tasks from this scenario assigned to Coy1 are written in BML as follows:

```
deploy BN-661 Coy1 at Denver end before TP0
   in-order-to enable label-o11 label-o10;
advance BN-661 Coy1 from Denver to Boston start
   at TP0 in-order-to enable label-o12 label-o11;
fix BN-661 Coy1 Red-1-182 at Boston end nlt TP1
   in-order-to enable label-o33 label-o12;
advance BN-661 Coy1 to Austin start at TP1
   in-order-to enable label-o14 label-o13;
fix BN-661 Coy1 Red-2-194 at Dog end nlt TP2
   in-order-to enable label-o35 label-o14;
advance BN-661 Coy1 to Atlanta start at TP2
   in-order-to enable label-o16 label-o15;
fix BN-661 Coy1 Red-2-196 at Atlanta end nlt TP3
   in-order-to enable label-o37 label-o16;
advance BN-661 Coy1 to Ruby start at TP3
   in-order-to enable label-o18 label-o17;
fix BN-661 Coy1 Red-2-191 at Lion end nlt TP4
```

```
   in-order-to enable label-o39 label-o18;
seize BN-661 Coy1 Lion at Lion end nlt TP4
   in-order-to cause label-ci1 label-o19;
```

Tasks assigned to Coy3 in BML are as follows:

```
deploy BN-661 Coy3 at Denver end before TP0
   in-order-to enable label-o32 label-o30;
support BN-661 Coy3 Coy1 at Troy start at TP0
   end at TP4 label-031;
attspt BN-661 Coy3 Red-1-182 from Denver to Boston
   start at TP0 end nlt TP1 in-order-to enable
   label-o12 label-o32;
destroy BN-661 Coy3 Red-1-182 at Boston end nlt TP1
   in-order-to enable label-o13 label-o33;
attspt BN-661 Coy3 Red-2-194 from Boston to Dog
   start at TP1 end nlt TP2 in-order-to enable
   label-o14 label-o34;
destroy BN-661 Coy3 Red-2-194 at Dog end nlt TP2
   in-order-to enable label-o15 label-o35;
attspt BN-661 Coy3 Red-2-196 from Dog to Atlanta
   start at TP2 end nlt TP3 in-order-to enable
   label-o16 label-o36;
destroy BN-661 Coy3 Red-2-196 at Atlanta end nlt TP3
   in-order-to enable label-o17 label-o37;
attspt BN-661 Coy3 Red-2-191 from Atlanta to Lion
   start at TP3 end nlt TP4 in-order-to enable
   label-o18 label-o38;
destroy BN-661 Coy3 Red-2-191 at Lion end nlt TP3
   in-order-to enable label-o19 label-o39;
```

10.7.2   Same Management Scenario in SGL

This scenario can be presented in SGL as follows.

```
FIXER:BN_661_Coy1;
SUPPORTER_DESTROYER:BN_661_Coy3;
deploy(Denver, T:TP0);
advance_destroy(
 (PL:Boston, TARGET:Red_1_182, T:TP1),
```

```
(PL:Austin, OBJ:DOG, TARGET:Red_2_194, T:TP2),
(PL:Atlanta, TARGET:Red_2_196, T:TP3),
(PL:Ruby, OBJ:LION, TARGET:Red_2_191, T:TP4));
seize(LION, T:TP4)
```

This semantic level description is much clearer, and more compact than if written in BML on the level of interacting individual units, as above. This simplicity may allow us redefine the whole scenario or its parts at runtime, on the fly, when the goals and environment change rapidly, also naturally engage robotic units instead of manned components. We may further represent this battlefield scenario at other levels, for example, moving upwards with its generalization, as follows:

- Not mentioning own forces, which may become clear at runtime only:

```
deploy(Denver, T:TP0);
advance_destroy(
 (PL:Boston, TARGET:Red_1_182, T:TP1),
 (PL:Austin, OBJ:DOG, TARGET:Red_2_194, T:TP2),
 (PL:Atlanta, TARGET:Red_2_196, T:TP3),
 (PL:Ruby, OBJ:LION, TARGET:Red_2_191, T:TP4));
seize(LION, T:TP4)
```

- Further up, not mentioning adversary's forces, which may not be known in advance but should be destroyed if discovered, to move ahead:

```
deploy(Denver, T:TP0);
advance(
 (PL:Boston, T:TP1),
 (PL:Austin, OBJ:DOG, T:TP2),
 (PL:Atlanta, T:TP3),
 (PL:Ruby, OBJ:LION, T:TP4));
seize(LION, T:TP4)
```

- Further up, setting main stages only, with starting and final time only known:

```
deploy(Denver, T:TP0);
advance(PL:(Boston, Austin, Atlanta, Ruby));
seize(LION, T:TP4)
```

- And final goal only (or just Command Intent, CI):

```
seize(LION, T:TP4)
```

Having the same formal language for all system levels and their any mixtures provides us with high flexibility for organization of advanced missions, especially with limited or undefined in advance resources and unknown environments, also possibility of potentially unlimited engagement of robotic components under unified command and control.

10.8   Distributed Avionics

The distributed control technology offered can be applied not only to systems distributed over certain physical spaces but also concentrated ones like, for example, aeroplanes (symbolically shown in Fig. 10.37) or satellites. Sometimes dealing with them in a fully distributed manner, without dedicated central resources, may prove to be beneficial, especially in unexpected or crisis situations [26].

SGT can help keep functional and control integrity of an object like plane even in most critical situations, say, when it disintegrates as a physical whole after being hit by bad weather, lightning, or missile. SGL interpreters can be implanted into key elements of the object and can communicate via wired or wireless links, forming altogether a sort of a spatial supercomputer which can operate without any central facilities (the latter being usually most vulnerable in crisis situations). Such organization may be keeping integrity of the object by self-spreading, self-replicating, and self-matching SGL scenarios dynamically creating any infrastructures needed.

Fig. 10.37 Distributed aircraft surveillance in SGL

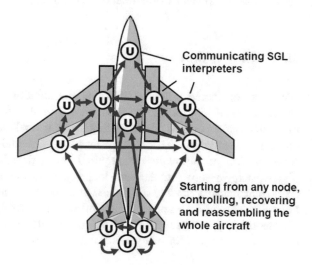

Communicating SGL interpreters

Starting from any node, controlling, recovering and reassembling the whole aircraft

We will show below only a very simple scenario which, starting in any point of the aircraft, can dynamically cover the whole object, verify states of its main mechanisms, collect results, and conclude on their sufficiency in comparison with the established repertoire registered in the Main database. The latter can be mobile too and not connected to any particular aircraft point in advance (which may accidentally happen to be out of operation, as any other point).

```
hop(random, any);
frontal(Main) =
   (left_aileron, right_aileron,
   left_elevator, right_elevator,
   rudder, left_engine, right_engine,
   left_chassis, right_chassis, ...);
repeat(
 nodal(Available) =
  repeat(
    free(verify(equipment); CONTENT),
    hop(first, neighbors));
  if(insufficient(Available, Main), set(alarm));
  hop(random, neighbors))
```

After performing global analysis of the whole aircraft by completely distributed navigation mechanisms, the scenario can freely (randomly in our case) move to other points and activate overall distributed and parallel analysis from them too, and so on. This provides constantly active and mobile supervision of the multi-component object, where its individual parts may happen to be out of order indiscriminately. More complex SGL scenarios can find, investigate, simulate, and implement different possibilities of restructuring and reconnecting of the remaining elements into another suitable operational whole.

10.9   Conclusion

We have considered different situations and solutions concerning command and control (C2) of distributed dynamic systems and campaigns. They ranged from fully automatic control, just resulting from formal networked interpretation of semantic scenarios in SGL, to explicit description and establishment of different types of C2 infrastructures, if required for application efficiency. The shown diversity expressed within the same language syntax and semantics may allow us to create and manage flexible distributed systems, manned, unmanned or mixed, for both civil and defense applications, which can be easily recovered after damages and reoriented in case of changing goals, situations, and environments. The exhibited examples for networked night vision, tracing moving objects with complex routes, or management of the whole missile defense system confirmed this. High-level holistic vision and description of distributed battlefields in SGL appeared to be much more transparent and compact in comparison to traditional parts-to-whole management approaches.

References

1. P. Sapaty, Integration of ISR with advanced command and control for critical mission applications, in *SMi's ISR Conference*, Holiday Inn Regents Park, London, United Kingdom, 7–8 April 2014
2. P.S. Sapaty. Global electronic dominance with spatial grasp. Int. J. Commun. Netw. Syst. Sci. (IJCNS). **5**(11) (2012)
3. P. Sapaty, Distributed capability for battlespace dominance, in *Electronic Warfare 2009 Conference and Exhibition*, Novotel London West Hotel & Conference Center, London, 14–15 May, 2009
4. P. Sapaty, Unified transition to cooperative unmanned systems under spatial grasp paradigm, in *19th International Command and Control Research and Technology Symposium*, Alexandria, Virginia, 16–19 June 2014
5. Sapaty, P, Remote control of open groups of remote sensors, in *SPIE Europe Security + Defence*, bcc Berliner Congress Centre, Berlin Germany (2009)
6. P.S. Sapaty, Advanced maritime operations under network-centric organizations, in *Proceedings of the International Conference Warship 2016: Advanced Technologies in Naval Design, Construction and Operation*, Guildhall, Bath, UK, 15–16 June 2016
7. Y. Xian, Cruise missile route planning based on quantum immune clone algorithm. J. Inf. Comput. Sci. **9**, 8 (2012), www.joics.com
8. K. Osborn, F-35 intercepts cruise missile to defend ship during important test, the BUZZ, http://nationalinterest.org/blog/the-buzz/f-35-intercepts-cruise-missile-defend-ship-during-important-17707
9. W.H. Zinger, J.A. Krill, Mountain top: beyond-the-horizon cruise missile defense, http://www.jhuapl.edu/techdigest/TD/td1804/zinger.pdf
10. JLENS: Co-ordinating cruise missile defense—and more, http://www.defenseindustrydaily. com/jlens-coordinating-cruise-missile-defense-and-more-02921/
11. J. Keller, Advanced military night-vision sensors rely on sensor fusion, networking, and signal processing, 1 January 2012, http://www.militaryaerospace.com
12. P. Sapaty, Night vision under advanced spatial intelligence: a key to battlefield dominance, in *International Summit Night Vision 2013*, London, 4–6 June 2013 http://www.smi-online.co. uk/defence/uk/night-vision
13. P.S. Sapaty, Distributed air and missile defense with spatial grasp technology. Int. J. Intell. Control Autom. (ICA), Sci. Res., **3**(2) (2012)
14. Ballistic missile defence, NATO, 2016, http://www.nato.int/cps/en/natolive/topics_49635.htm
15. NATO missile defence system, Wikipedia, https://en.wikipedia.org/wiki/NATO_missile_ defence_system
16. P.S. Sapaty, High-level management of distributed air and missile defence systems, in *International Summit Air and Missile Defence India*, The Taj Mahal Hotel, New Delhi, India, 5–6 June 2012
17. P. Sapaty, in Distributed Air and Missile Defense with Spatial Grasp Technology (International Summit Military Space, London) 10–11 April 2013
18. P. Sapaty, Air and missile defense with spatial grasp technology, in *Proceedings of Unmanned Systems Asia 2011*, Pre-Conference Workshop B, Singapore 2011

19. P.S. Sapaty, Distributed missile defence with spatial grasp technology, in *International Conference Military Space*, Holiday Inn Regents Park London, 4–5th March 2015
20. P. Sapaty, High-level organisation and management of directed energy systems, in *Proceedings of Directed Energy Weapons 2010*, CCT, Canary Wharf, London UK, 25–26 March 2010
21. U. Schade, M.R. Hieb, Formalizing battle management language: a grammar for specifying orders. 06S-SIW-068, presented at the 2006 spring simulation interoperability workshop, April 2006, Huntsville, AL
22. U. Schade, M. R. Hieb, M. Frey, K. Rein, Command and control lexical grammar (C2LG) specification, FKIE Technical Report ITF/2010/02, July 2010. http://c4i.gmu.edu/eventsInfo/conferences/2011/BMLsymposium2011/papers/BML-Symposium-Schade.pdf
23. Coalition Battle Management Language (C-BML), RTO Technical Report TR-MSG-048, February 2012., http://www.dtic.mil/dtic/tr/fulltext/u2/a559456.pd
24. M. Powers, M. Hieb, M. Pullen, M. Kleiner, A geospatial battle management language (GeoBML) for terrain reasoning, in *11th International Command and Control Research and Technology Symposium* 2005. http://www.dodccrp.org/events/11th_ICCRTS/html/presentations/110.pdf
25. P.S. Sapaty, Formalizing commander's intent by spatial grasp technology, in *2012 Annual Conference accepted paper at the International Society of Military Sciences (ISMS)*, Kingston, Ontario, Canada, 23–24 October 2012. http://www.isofms.org/cms_uploads/Sapaty_Abstrac2012t.pdf
26. P. Sapaty, Grasping the whole by spatial intelligence: a higher level for distributed avionics, in *Proceedings of International Conference Military Avionics 2008*, Café Royal, London, UK, Jan 30–Feb 1

Chapter 11
Collective Robotics

11.1 Introduction

Large numbers of robotic facilities have been accumulated worldwide in almost every sphere of human activity, and their further development is taking place with increased financing and speed. A brief demonstration of advanced examples of ground, aerial and maritime robots, also becoming popular their collective applications will be provided. But in many cases the developed robots still remain specialized devices rather than intelligent collaborators for humans, often having dissimilar and proprietary architectures and specific management and control.

To embed advanced robotics seamlessly into human societies at even greater scale, much more general and universal approaches need to be developed for their tasking and autonomous decision-making, also effective integration with human C2 infrastructures. In this respect, a number of collective robotic scenarios will be considered in SGL operating at different organizational levels and their mixtures under unified control provided by automatic SGL interpretation.

These combine randomized swarm movement with regularly updated hierarchical C2 optimized at run time for collecting and impacting distributed targets, show the use of multiple robots in hospital service, which are moving via corridors between hospital rooms avoiding collisions and providing the assistance needed, also describe robotic mapping of unknown spaces with subsequent use of the obtained map for optimal movement through the territory.

Other scenarios exhibit battling forest fires by robotic swarms after outlining disaster zones, unmanned costal waters collective patrol reporting the unwelcomed objects if discovered, collective finding of oil spill centre in water covered areas, massive robotic attack on adversary's fleet in maritime environment, swarm against swarm collective aerial operation reacting on the frequenting malicious use of unmanned aerial vehicles, and engagement of multiple drones in forestry and agriculture.

© Springer International Publishing AG 2017
P.S. Sapaty, *Managing Distributed Dynamic Systems with Spatial Grasp Technology*, Studies in Computational Intelligence 690,
DOI 10.1007/978-3-319-50461-2_11

11.2   Some Modern Robotic Examples

11.2.1   Ground Robotics

The ability of robots to save lives has secured future path for ground robotics alongside the warfighter. Ground robotics can be engaged in different missions including Explosive Ordnance Disposal (EOD), Combat Engineering, Reconnaissance, and many others. The Robots designed by Boston Dynamics may be named among the latest developments in ground robotics, see Fig. 11.1.

The LS3 "*robot mules*", as in Fig. 11.1a, are to help soldiers carry heavy loads [1]. It is a rough-terrain robot designed to go anywhere marines and soldiers go on foot, helping carry their load. Each LS3 carries up to 400 lbs of gear and enough fuel for a 20-mile mission lasting 24 h. It automatically follows its leader using computer vision, so does not need a dedicated driver, also travels to designated locations using terrain sensing and GPS.

The *Cheetah robot*, its concept in Fig. 11.1b, is the fastest legged robot in the world, surpassing 29 mph [2]. The Cheetah robot has an articulated back that flexes back and forth on each step, increasing its stride and running speed, like animals do. Its current version runs on a high-speed treadmill in the laboratory being powered by an off-board hydraulic pump.

The next generation Cheetah robot, *WildCat*, Fig. 11.1c, is designed to operate untethered. It is funded by the Defense Advanced Research Projects Agency (DARPA) and is being developed for military use. With a large motor attached, WildCat isn't as fast as its cousin, being currently limited to around 16 mph on flat terrain.

11.2.2   Aerial Robotics

Unmanned flying vehicles (UAVs), like the ground vehicles, have dual applications: they can be used for reconnaissance without endangering human pilots, and also can carry missiles and other weapons [3, 4]. Among the best known armed

(a)   **(b)**   **(c)**

Fig. 11.1 Boston dynamics robots. **a** Robot mules; **b** the cheetah concept; **c** cheetah becoming wild cat running untethered

(a)   **(b)**   **(c)**

Fig. 11.2 Aerial robots. **a** Northrop Grumman X-47B; **b** SR-72 with Mach 6; **c** DARPA HTV-2 with Mach 20

UAVs are the semi-autonomous Predator Unmanned Combat Air Vehicles (UCAV) built by General Atomics which can be equipped with Hellfire missiles. The military services are also developing very small aircraft, sometimes called Micro Air Vehicles (MAV) capable of carrying a camera and sending images back to their base. Some newest UCAV developments are mentioned below.

The Northrop Grumman X-47B, as in Fig. 11.2a, is designed for carrier-based operations [5]. Its project began as part of DARPA's J-UCAS program, and is now part of the Unmanned Combat Air System Demonstration (UCAS-D) program.

Northrop Grumman intends to develop the prototype X-47B into a battlefield-ready aircraft, which will enter service around 2019. X-47B can stay in the air for 50 h, carry 2 tons of weaponry, and be refuelled in the air.

Hypersonic aircraft. The SR-72 [6], Fig. 11.2b, could fly as fast as Mach 6, will have the ability to gather intelligence, conduct surveillance and reconnaissance, and launch combat strikes at an unprecedented speed. SR-72 could be operational by 2030. At this speed the aircraft would be so fast that adversary would have no time to react.

DARPA rocket-launched HTV-2, Fig. 11.2c, 13,000 mph *Hypersonic Glider* [7], was designed to collect data on such challenges of hypersonic flight as aerodynamics, aerothermal effects, and guidance, navigation and control. The HTV-2's test flight was conducted to validate current models and increase technical understanding of the hypersonic regime, which successfully demonstrated stable aerodynamically-controlled flight at speeds up to Mach 20.

11.2.3 Maritime Robotics

Sea-based robots as Unmanned Maritime Systems, or UMSs, can be either free-swimming or tethered to a surface vessel, a submarine, or a larger robot [8], see examples in Fig. 11.3.

Tethers simplify providing power, control, and data transmission, but limit manoeuvrability and range. Recently highly autonomous systems have been built that can navigate, manoeuvre, and carry out complex tasks. UMSs can operate on

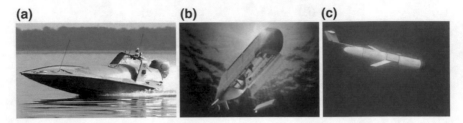

Fig. 11.3 Maritime robots. **a** Unmanned surface vehicle; **b** large displacement unmanned undersea vehicle, LDUUV; **c** underwater glider

the ocean's surface, just below the surface, or entirely underwater. The unmanned surface vehicles (USVs), Fig. 11.3a, are already widely used in practice.

Large Displacement Unmanned Undersea Vehicle (LDUUV) [9], see Fig. 11.3b, is to conduct missions longer than 70 days in open ocean and littoral seas, being fully autonomous, long-endurance, land-launched, with advanced sensing for littoral environments. It is expected to reach initial operating capability by 2020 and full rate production by 2025.

Underwater gliders [10], see Fig. 11.3c, will use a process called "hydraulic buoyancy," which allows the drone to move up and down and in and out of underwater currents which provides the speed of about one mile per hour. Carrying a wide variety of sensors, they can be programmed to patrol for weeks at a time, surfacing to transmit their data to shore while downloading new instructions at regular intervals.

11.2.4   Collectively Behaving Robots

To be of real help in complex civil and military applications, robots should be integral part of manned systems, they should also be capable of operating massively, in large robotic collectives, with some examples shown in Fig. 11.4.

Fig. 11.4 Massive use of robotic units. **a** Tests on Virginia's James River; **b** Harvard University multiple robots operating without central intelligence; **c** sci-fi image of future robotic armies

The tests on Virginia's James River represented the first large-scale military demonstration of a *swarm of autonomous boats* designed to overwhelm enemies [11], see Fig. 11.4a. The boats operated without any direct human control; they acted as a robot boat swarm.

Harvard University scientists have devised a *swarm* of 1024 tiny robots that can work together without any guiding central intelligence [12], see Fig. 11.4b. These robots can assemble themselves into five-pointed stars, letters of the alphabet and other complex designs. Swarm scientists are inspired by nature's team players like bees, ants and termites, schools of fish, and flocks of birds. These creatures collaborate in vast numbers to perform complicated tasks, even though no single individual is actually in charge.

These results are believed to be useful for the development of advanced robotic teams, even armies, with some futuristic image in Fig. 11.4c.

11.2.5 General Demands to Advanced Robotic Systems

A thorough analysis of aims and results of the development and implementation of both civil military robots, including and especially those briefed above, helps us formulate some general features and demands in relation to their overall management and control, as follows:

- Despite the diversity of sizes, shapes, and orientations, they should all be capable of operating in distributed often large physical spaces, thus falling into the category of distributed systems.
- Their activity is to include navigation, movement, observation, gathering data, carrying loads which may include ammunitions or weapons, and making impact on other manned on unmanned units and the environment.
- They should have certain, often high, degree of autonomy and capability of automatic decision making to be really useful in situations where human access and activity are restricted.
- They should effectively interact with manned components of the systems and operate within existing command and control infrastructures, to be integral parts of the system.
- They should be capable of effective swarming for massive use, and this swarming should be strongly controlled from outside—from manned parts of the system or from other, higher-level, unmanned units.
- Their tasking and re-tasking (including that of swarms) should be flexible and convenient to humans to guarantee runtime reaction on changing goals and environments, especially on battlefields.
- The use of unmanned units should be safe enough to humans and systems they are engaged in.
- Their behaviour should satisfy ethical and international norms, especially in life-death situations.

The following examples of organization of robotic systems with the use of SGT are prepared within the context of listed above demands, many of which can be effectively satisfied under the technology developed.

11.3   Integration of Loose Swarming with Hierarchical Command and Control

Imagine that a distributed area needs to be investigated by multiple unmanned aerial vehicles, as in Fig. 11.5, which should randomly search the space, collect information on unwanted objects, classifying them as targets, and organize collective reaction on emerging threats.

The initial distribution of the aerial group is shown in Fig. 11.6, and different group functionalities can be expressed in SGL as follows.

- Swarm movement scenario, let us call this **swarm_move**. Randomized swarm movement of individual robots, starting in any node and with minimum, threshold Range between moving nodes allowed.

```
#all;
nodal(
  Limits = (dx(0,8), dy(-2,5)), Range = 200, Shift);
  repeat(
```

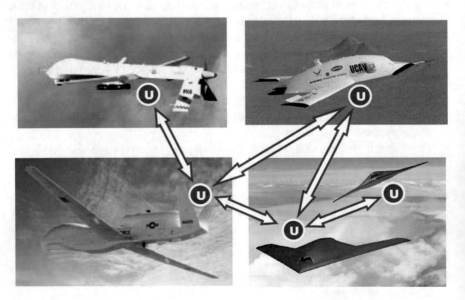

Fig. 11.5 UAV and UCAV examples with communicating SGL interpreters embedded

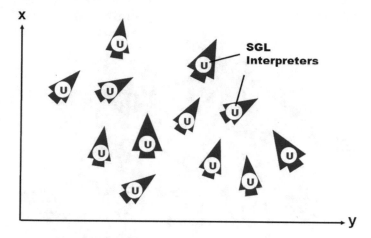

Fig. 11.6 Initial distribution of robotic units

```
Shift = random(Limits);
if(empty(Shift, Range), WHERE += Shift);
sleep(delay))
```

A snapshot of such swarm movement is shown Fig. 11.7.

- Finding topologically central unit and hopping into it, starting from any unit, calling it **find_hop_center**, see Fig. 11.8.

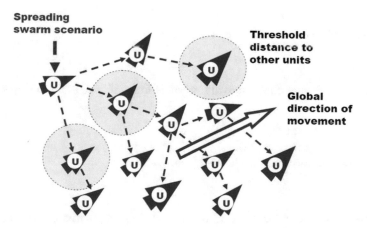

Fig. 11.7 Swarm movement snapshot

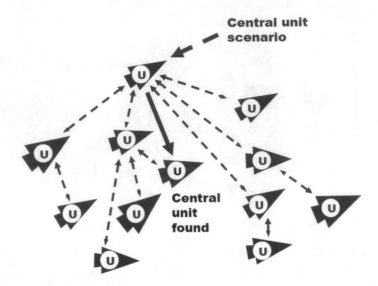

Fig. 11.8 Finding central unit

```
frontal(Aver) = average(#all; WHERE);
min_destination(#all; distance(Aver, WHERE))
```

The group's center found by this scenario may be as in Fig. 11.8.

- Creating runtime infrastructure starting from the central unit found using oriented semantic links `infra` and maximum allowed physical distance or `Depth`
  between nodes to form direct links in between, as follows (see Fig. 11.9), and let
  this be named **infra_build**.

```
frontal(Depth = 1000);
repeat(linkup(+infra, firstcome, Depth))
```

With more details, first removing previous links, if any, at the nodes visited:

```
frontal(Depth = 1000);
repeat(#(firstcome, Depth);
    stay_unlink(all); linkup(-infra, BACK))
```

- Targets collection and distribution and impact (naming this as **col-
  lect_distribute_impact**) starting from the central unit found, as in
  Fig. 11.10.

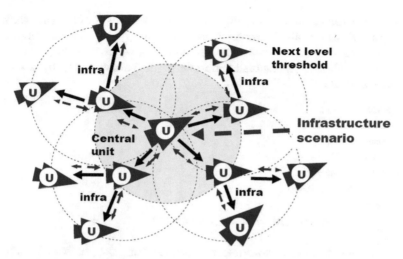

Fig. 11.9 Runtime infrastructure creation

Fig. 11.10 Targets collection, dissemination, and impact

```
frontal(Seen) =
  repeat(free_detect(targets), +infra#);
nonempty(Seen);
repeat(free_select_shoot(Seen), +infra#)
```

- **Scenarios integration**. Using the above SGL scenarios for different behavioral stages, we can easily integrate them within the global one, as follows, by substituting their names with corresponding full scenario texts (where *time* stands for allowed operational time interval to be substituted by a concrete numerical value):

```
#(random, all);
parallel(
 (swarm_move),
 repeat(
  find_hop_center;
  stay(infra_build);
  allowed(time,
      sling(collect_distribute_impact)))))
```

The above mentioned constituent scenarios can also be represented and remembered as procedures and called by their names instead of placing their full bodies into the resultant integral scenario.

The obtained resultant scenario combines loose, randomized swarm movement in a distributed space with periodic updating of topologically central unit (as units are changing distances and relative positions) and updating runtime hierarchical infrastructure between them. This infrastructure controls observation of distributed territory while collecting potential targets, distributing them back to the vehicles for local assessment, selection and impact.

11.4 Multi-robot Hospital Service Example

Robots are now widely used in hospitals too [13], as shown in Fig. 11.11.

A possible multi-robot service scenario is depicted in Fig. 11.12, where robots are randomly moving between six hospital units, checking emergent demands and providing services needed (while avoiding collisions with each other).

Fig. 11.11 Hospital robotic service

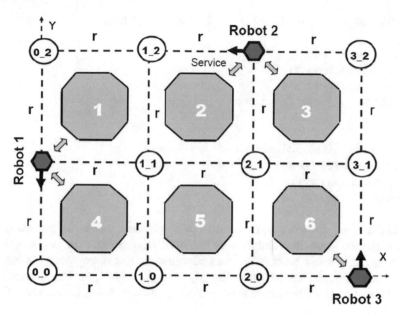

Fig. 11.12 Multi-robot hospital service organization

Creation of a distributed campus map, to be subsequently used by the service robots, may be as follows (starting from point 3_1 and using depth-first spanning tree template, see also Chap. 7):

```
create(
 #3_1; r#2_1; r#1_1; r#0_1;
 (r#0_2; r#1_2; r##1_1,
  (r#2_2; r##2_1,(r#3_2; r##3_1))),
 (r#0_0; r#1_0; r##1_1,
  (r#2 0; r##2 1,(r#3 0; r##3 1))))
```

Simultaneous random movement with the help of this map and the service needed can be expressed as follows (with collision avoidance, also assuming that robots, if available and chosen, are placed initially into positions 0_1, 2_2, and 3_0):

```
#(0_1, 2_2, 3_0);
nodal(Mark = 1); WHERE = NAME;
repeat(
 or(
  ((random, all)#; seize(Mark == nil; Mark = 1);
  WHERE = NAME; (#BACK; Mark) = 0;
  perform(service)),
  stay))
```

With assigning particular robots explicitly, we will have:

```
(DOER = Robot1; #0_1),
(DOER = Robot2; #2_2),
(DOER = Robot3; #3_3);
nodal(Mark = 1); WHERE = NAME;
repeat(
 or(
  ((random, all)#; size(Mark == nil; Mark = 1);
  WHERE = NAME; (#BACK; Mark) = 0;
  perform(service)),
  stay))
```

This scenario can work with any number of available robots, where new robots can be injected at any time and from any points on the created map (not occupied at this moment by other robots). We may also assume that some points are hospital entry ones, say, at the left side like 0_0, 0_1, and 0_2. The scenario for involvement of new robots from these entries (randomly chosen for each robot) will be as follows:

```
nodal(Entry = (0_0, 0_1, 0_2), Mark);
sling(
 #(random, Entry); seize(Mark == nil; Mark = 1);
 WHERE = NAME;
 repeat(
 or(
  ((random, all)#; size(Mark == nil; Mark = 1);
  WHERE = NAME; (#BACK; Mark) = 0;
  perform(service)),
  stay)))
```

11.5 Exploration and Mapping of Unknown Distributed Space

Some preliminary research in this area was reported at [14]. Imagine there is an unknown distributed space. Starting from any point, it is needed to create a map of its free access part, i.e. where it is possible to move freely from any point to any other one. Using this map, it will be easy to organize optimized movement between any two physical points. An example of space to be searched, starting from some initial point, is shown in Fig. 11.13.

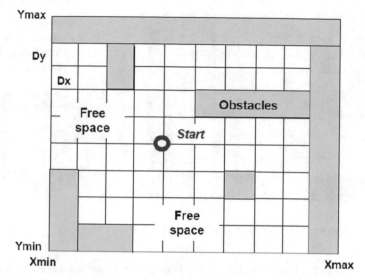

Fig. 11.13 Space to be mapped from a starting point

For simplicity, it is assumed that the space will be searched and mapped according to a square grid with given coordinate limits (Xmin, Xmax, Ymin, Ymax) and navigation steps being same on both coordinates (i.e. Dx == Dy or merely D). Grid node names will be represented as aggregates of two coordinate values, like X_Y.

11.5.1 Different Mapping Scenarios

- **Sequential solution based on depth-first spanning tree**, as in Fig. 11.14, starting from any physical location, will be as follows:

```
frontal(Xmin = ..., Xmax = ..., Ymin = ..., Ymax = ...,
    X, Y, D =..., Cover);
assignpeers((X, Y), WHERE(1, 2)); create(X_Y);
Cover =
{sequence(
  X += D, X -= D, Y += D, Y -= D;
  Xmin <= X <= Xmax; Ymin <= Y <= Ymax;
  nonequal(X_Y, PREDECESSOR);
  or(blind_linkup_noback(r, X_Y), create(r, X_Y));
  WHERE = (X, Y); apply(Cover))};
apply(Cover)
```

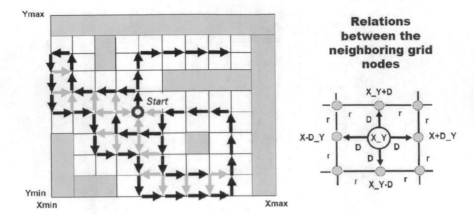

Fig. 11.14 Space mapping based on depth-first spanning tree

It is necessary to express in SGL only forward robotic movement, whereas its return to the previous physical positions of the growing depth-first spanning tree is organized fully automatically by SGL interpreter with the support of its track system, which nodes are associated with the already created grid nodes.

As only a single robot is engaged, and the search is based on depth-first spanning tree, there can be no conflicts in creating grid nodes and links.

- **Parallel solution with breadth-first spanning tree**, see Fig. 11.15, can be organized by the following SGL scenario, also starting from a single physical point.

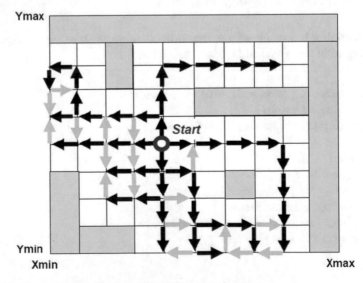

Fig. 11.15 Space mapping based on breadth-first spanning tree

The general SGL solution for this case will look like follows:

```
frontal(Xmin = ..., Xmax = ..., Ymin = ..., Ymax = ...,
    X, Y, D =...);
assignpeers((X, Y), WHERE(1, 2)); create(X_Y);
repeat(
X += D, X -= D, Y += D, Y -= D;
Xmin <= X <= Xmax; Ymin <= Y <= Ymax;
nonequal(X_Y, PREDECESSOR);
or(blind_linkup_noback(r, X_Y),
   (empty_location(X_Y); create(r, X_Y)));
WHERE = X_Y)
```

As the space coverage evolves in parallel, possible conflicts in creating the same
new map nodes from simultaneous branches can be resolved by observing neighboring physical space by sensors in individual robots, and blocking their simultaneous node creation and movement into the same locations. With a totally unknown
environment, the robots should preferably start as a located group in the starting
point for subsequent space navigation, but any other possibilities can be readily
supported in SGL too. Multiple cooperating robots can be engaged for this scenario,
see Fig. 11.16. For particular robots to be engaged, the latest scenario may start
with their explicit naming, as follows:

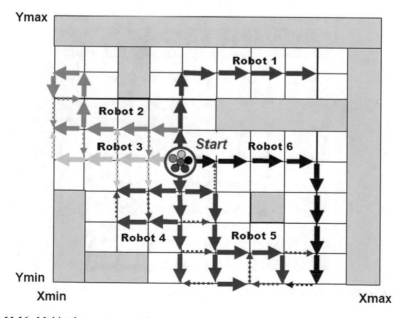

Fig. 11.16 Multi-robot space mapping

```
RESOURCES =
(Robot1, Robot2, Robot3, Robot4, Robot5, Robot6)
```

If the number of available robots is smaller than the number of fringe spanning tree nodes, some robots will have to move back to start serving other branches, with their safe return is guaranteed and organized automatically by the SGL interpreter (similar to the previous depth-first coverage example).

11.5.2   Finding Optimal Route by the Created Free Space Grid

The created map (as grid) of the free space can be distributed between the robotic units or assembled in one piece within a single memory. This map can also contain with its nodes some additional details of the related space locations, which can be investigated, measured, and recorded during the space navigation.

This map can be effectively used for many purposes, for example, finding and following optimal paths (like shortest ones) between any two nodes, let them be A and B, as in Fig. 11.17, where given exact coordinates of A and B may not fully correspond to the grid node coordinates in the map created.

The related SGL scenario for finding optimal path from A to B and subsequently moving via it physically may be as follows:

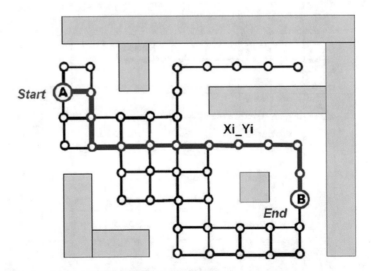

Fig. 11.17 Finding optimum route via the map created

```
frontal(Acoord = ..., Bcoord = ..., Far);
nodal(Distance, Before);
stay(
 (min_destination(#all; distance(Bcoord, NAME)));
 Distance = 0;
 repeat(all#; Far += 1;
     or(Distance == nil, Distance > Far);
     Distance = Far; Before = PREVIOUS)));
min_destination(#all; distance(Acoord, NAME));
repeat(WHERE = NAME; #Before); WHERE = Bcoord
```

Staying initially in physical position A, the scenario above first creates the shortest path tree from the grid node closest to B and covering the whole map. It then organizes physical movement from node A to the nearest grid position, then stepwise to the tree's root, and finally to exact coordinates of B.

11.6   Battling Forest Fires with Robotic Swarms

Fighting forest fires is a complex problem, and the use of drones may prove very effective [15]. We will consider a solution where distributed physical space is randomly searched by simultaneous propagation of multiple reconnaissance units, which when discover irregularities (e.g. forest fires) move further and encircle the fire zones, collect their perimeter coordinates, transfer them to some headquarters (HQ), and ultimately initiate massive impact on the zones under fire. The zones with fires and initial positions of reconnaissance units are shown in Fig. 11.18a, and positions of unites after they started moving randomly within certain coordinate sector are in Fig. 11.18b.

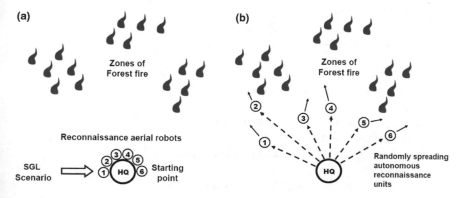

Fig. 11.18 Battling a forest fire. **a** Initial scenario injection; **b** robots movement

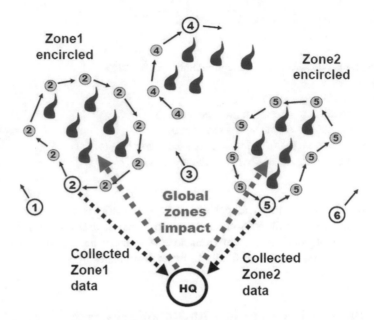

Fig. 11.19 Encircling fire zones followed by global impact

After detecting fire locations, the reconnaissance units that reached them begin moving around the fire zones, randomly choosing encirclement orientation (i.e. clockwise or anticlockwise). In each step they accumulate coordinates of the periphery of fire zones, and upon termination of the encirclement send the completed zone coordinates to HQ. Getting the latter, the HQ is launching a massive direct impact on the zones outlined, as in Fig. 11.19, which may be manned, unmanned, or mixed.

The full SGL scenario for this task may be as follows:

```
move(HQ); create_nodes(1, 2, 3, 4, 5, 6);
frontal(Zone, Limits = ..., Direction, Depth = ...,
     Threshold = ...);
repeat(
 shift_random(Limits);
 if(check(fire),
   (Zone = WHERE;
   Direction = random(clockwise, anticlockwise);
   repeat(
    move_around(fire, Direction, Depth);
    append(Zone, WHERE);
    if(distance(WHERE, Zone[1]) < Threshold,
      (hop_direct(HQ);
      blind_impact(Zone, global))))))))
```

Other interpretations of this scenario may be dealing with zones of industrial pollution, zones of fish concentration, zones of social unrest, etc., with aerial, ground, surface or underwater robots engaged.

11.7   Coastal Waters Cooperative Patrol

The scenario is as follows. A number of coastal patrol vehicles, which may be surface or underwater, are patrolling simultaneously the coastline, following it and regularly reporting of what they see or discover (sensors dependent) in key points.

- **Coastal map creation**. At the beginning we should create discrete coastal map as a semantic network consisting of coordinates of key points linked with each other by oriented links, as in Fig. 11.20.

To create this network, the following scenario is sufficient (all links assumed having same name r):

```
create(#x1_y1; +r#x2_y2; +r#x3_y3; ... +r#x9_y9)
```

This network can be stored in different ways, with all its nodes in a single world point up to each node in a separate point, the latter, say, corresponding to the physical locations of Fig. 11.20, assuming they can communicate with each other.

- **Single vehicle solution**. A single USV (or UUV) solution repeatedly navigating all coastal area by the map created is shown in Fig. 11.21 and SGL scenario that follows (searching the water space for unwanted objects with the depth available by vehicle's sensors).

```
frontal(Direction = +r, SearchDepth = ...);
#x1_y1;
repeat(
 repeat(WHERE = CONTENT;
```

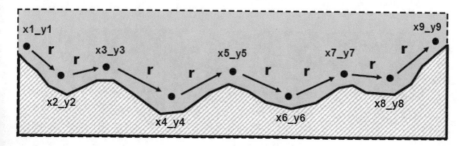

Fig. 11.20 Coastal waypoint map

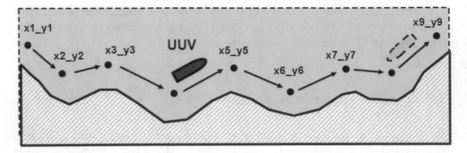

Fig. 11.21 Patrolling coastal waters with a single vehicle

```
        report_check(SearchDepth, alien);
        Direction#);
    invert(Direction))
```

- **Two vehicles solution**. Let us assume that two vehicles UUV1 and UUV2 are engaged, as in Fig. 11.22, which start from the opposite ends, following the whole route but in different directions.

  Two variants are considered below.

1. Vehicles moving forward and backward independently by the coastal map, avoiding collisions with each other (the avoidance mechanisms are supposedly present at implementation level), and report what they see/sense when pass the key points xi_yi.

```
frontal(Direction, SearchDepth = ...);
branch(
 (#x1_y1; Direction = +r),
 (#x9_y9; Direction = -r);
 repeat(
  repeat(WHERE = CONTENT;
       report_check_all(SearchDepth, alien);
       Direction#);
  invert(Direction)))
```

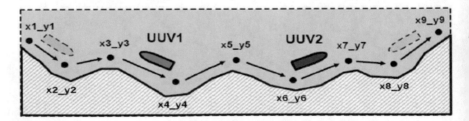

Fig. 11.22 Simultaneous coastal patrol

Each vehicle is synchronously moving in both virtual and physical spaces, with the next physical coordinates picked up from the corresponding semantic network nodes. Having reached the end of the virtual chain, they begin moving in the opposite direction through it unless reach the opposite end, and so on.

2. This solution differs from the previous one in that each vehicle turns back when discovers another vehicle in a threshold Distance, which means that the path ahead has been checked already (or will be) by the "colleague":

```
frontal(Direction = +r, SearchDepth = ...,
      Distance = ...);
branch(
 (#x1_y1; Direction = +r),
 (#x9_y9; Direction = -r);
 repeat(
 repeat(WHERE = CONTENT;
       report_check_all(SearchDepth, alien);
       none(colleague, Distance); Direction#);
 invert(Direction)))
```

The solutions above can be extended to any number of robotic units acting cooperatively.

11.8 Cooperative Finding of Oil Spill Center

This scenario describes the situation where a hypothetical region of the sea, or *expected_spill_area* (may be both surface and submersed) is polluted with spilling oil, and a group of communicating surface or underwater (or both) robots, initially delivered to the approximate region, is trying to detect the spill center cooperatively, reporting the finding if successful, as in Fig. 11.23.

The following SGL scenario may start from any robot, eventually covering the whole group via regular communications between the units. Each robot tries to move stepwise toward increased spill level in a randomized way (within *move_range*), otherwise returns to the previous position if the move was unsuccessful. Each movement is allowed only if other robots remain in reach (by *communication_range*), to maintain group's connectivity as a system.

Fig. 11.23 Collective finding
of oil spill center

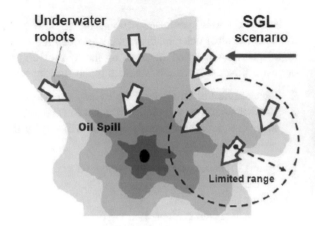

```
#all;
move(random, expected_spill_area);
nodal(Level, Direction = 0, Current, Count, Max);
Level = check(spill);
branch(
 loop(
  Max = maximum(
   (#(communication_range, all); Max), Max, Level);
  If(Level == Max, increment(Count));
  if(Count >= tests_threshold_number,
    stop_output('Center:', WHERE));
  sleep(Delay)),
 loop(
  Current = WHERE;
  or((WHERE += random(move_range, Direction);
    nonempty(#(communication_range, all));
    New = check(spill) > Level;
    Level = New; Count = 0;
    Direction = angle(Current, WHERE)),
   WHERE = Current)))
```

By regular local communications between robots, the currently maximum spill
level discovered by the group (Max) as a whole is constantly updated and maintained in each robot. And the robot(s) where this maximum level corresponds to the
local spill level after a threshold delay (expressed by a number of repeated tests, or
`tests_threshold_number`) report the spill's center (which may be submersed too).

11.9   Maritime Massive Robotic Attack

This scenario relates to the use of a large unmanned swarm for highly dynamic
distributed operations, where robotic units are supposed to be naval, surface ones,
as in [16, 17], and the operation is a hypothetical swarm attack on an adversary's
fleet. SGL can describe and provide goal-driven behaviour of the robotic swarm
which can operate autonomously and without any central resources. The initial
stage of the scenario is depicted in Fig. 11.24, with the swarm being outside the
expected attack area.

The following SGL high-level (generalized, without details) scenario can start
from any robot while covering the whole swarm and causing units to move stepwise
to the area by expected *area_coordinates*, with each planned step using
next_step_guidelines and *allowed_distance* between robotic units.

```
#all;
nodal(Area     = area_coordinates,
    Guidelines = next_step_guidelines,
    Range1     = allowed_distance,
    Range2     = vision_range,
    Range3     = shooting_range,
    Range4     = communication_range,
    Offer);
frontal(Transit);
sling(
```

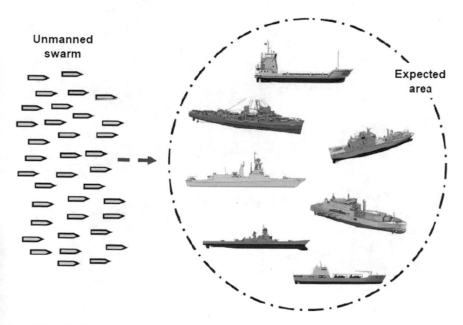

Fig. 11.24 Initial stage of robotic attack

```
Offer = propose(WHERE, Area, Guidelines);
if(none(Offer, Range1), WHERE += Offer);
branch(
 (correct(Area, detect(Range2, all)));
 Transit = Area; #(Range4, all);
 update(Area, Transit)),
 impact_detect(targets, Range3)))
```

Each unit constantly updates the initially given approximate region coordinates by what it currently sees within the *vision_range*, also regularly sharing its own improved area version with the neighboring robotic units staying within *communication_range* (regularly updated from the neighbors too, and thus throughout the entire swarm).

The whole swarm in such cooperative way is gradually improving global image of the expected area, and in each robot. With continually improved area coordinates, the next movement steps are aimed at surrounding of the area rather than moving towards its center in order to prevent possible escape from the attack.

In parallel with this collective move towards the target vessels located within the attack area and nearing them (with the area coordinates updates helping, in turn, to correct the very moves) another process is trying to detect the adversary's units within *shooting_range* from the robotic units and attack these targets. A possible development snapshot for this scenario is shown in Fig. 11.25.

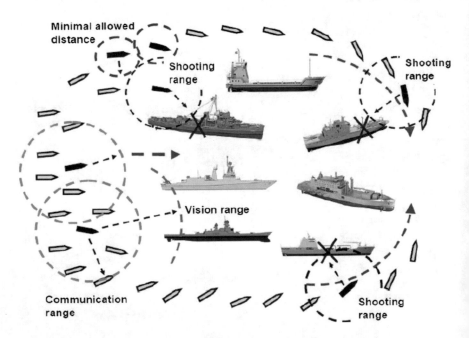

Fig. 11.25 Subsequent stages: swarm coordinated movement with target attacks

11.10 Swarm Against Swarm Aerial Scenario

We will consider here the case where an unmanned swarm is opposing another, supposedly unwanted, group of aerial vehicles. This, for example, can relate to fighting criminal and spying drones which are currently spreading worldwide [18, 19], and may potentially operate in swarms too (some symbolic drone images are shown in Fig. 11.26).

Main ideas of the following swarm against swarm SGL scenario, with alien drones as Targets and friendly drones as Chasers are shown in Fig. 11.27.

```
frontal(Chasers = ..., Targets, Next, List, Center);
repeat(
 #(random, Chasers);
```

Fig. 11.26 Drones and their symbolic swarms to fight

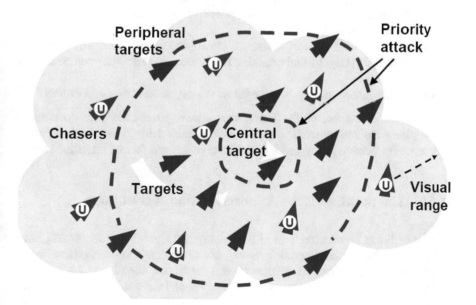

Fig. 11.27 Fighting group targets with unmanned swarms

```
Targets = merge(#Chasers; coordinates(visible));
nonempty(Targets); Center = average(Targets);
List = sortdown(
 split(Targets); distance(VALUE, Center) & VALUE);
List = append(withdraw(List, last), List);
sling(
  nonempty(List); Next = withdraw(List, 1) : 2;
  min_destination(
    #Chasers; STATUS == vacant;
    distance(WHERE, Next));
  STATUS = engaged;
  free(pursue_neutralize(Next);
     STATUS = vacant)))
```

Main steps and details of this scenario:

1. Initial launch of the swarmed chasers (in Fig. 11.27 with SGL interpreters U embedded, which can communicate with each other) into the expected operational area.
2. Discovering targets and forming priority list by their physical positions where maximum priority is assigned to topologically central targets as potential control units of the intruders.
3. Other targets are sorted by their distance from the topological center of their group.
4. The most peripheral targets (those in maximum distance from the topological group's center), are assigned higher priority too as potentially having higher chances to escape.
5. Assigning available chasers to targets, classifying them as engaged, chasing and neutralizing targets, and returning into status vacant after performing the task.
6. The vacant chasers are again engaged in the targets selection and impact.

It is worth noting that all the chaser swarm management has been done exclusively within the swarm itself, without external intervention, which can dramatically simplify outside group tasking and involve any number of unmanned units.

11.11   Cooperative Robotic Forestry and Agriculture

Aerial robots can be effectively used for forestry and agriculture too [20–22], from aerial observation to planting to harvesting, where efficient cost-competitive fleet of drones can drastically reduce reliance on manual labour, see Fig. 11.28.

We will be considering here some scenarios of land navigation in a regular manner and doing something in the points reached. This may relate to spreading fertilizers, dropping seeds, or picking certain crops from the air, with the used area

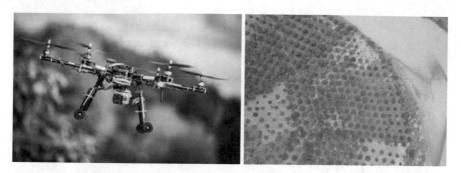

Fig. 11.28 Using drones in forestry and agriculture operations

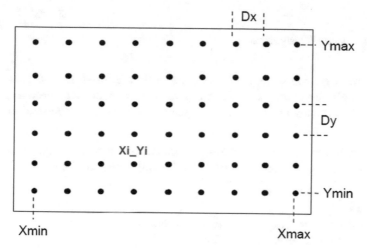

Fig. 11.29 The area parameters

parameters shown in Fig. 11.29. (Will be mentioning seeds dropping only, but any other actions can be used too in the same scenarios.)

The following variants are considered with different description levels and different numbers of available drones for their execution.

- **Top semantic level, just intent** in SGL, where the task is formulated on most general level, just where and what to do, without any details of availability of execution drones, their number, also their organization as a system.

```
drop(seeds,
    area(Xmin =...; Xmax =...; Ymin =...; Ymax =...,
    Dx =...; Dy =...))
```

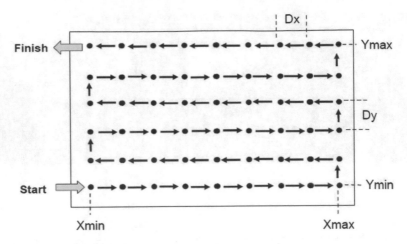

Fig. 11.30 Sequential navigation

This most general scenario can be executed in a great variety of ways, starting with a single drone and up to as many of them as there are separate points in Fig. 11.29, with any order of their engagement and any paths followed to perform the jobs.

- **Sequential territory navigation**, horizontal, left-to-right then right to left and so on, see Fig. 11.30, for which the following scenario will be sufficient.

```
frontal(
  X = Xmin = ..., Xmax = ..., Y = Ymin = ..., Ymax = ...,
  Dx = ..., Dy = ..., Direction = right);
move(X, Y);
repeat(
  or(
   (Direction == right;
    if(X + Dx <= Xmax, X += Dx,
      (Y += Dy; Direction = left))),
   (Direction == left;
    if(X - Dx >= Xmin, X -= DX,
      (Y += Dy; Direction = right))));
  Y <= Ymax;
  move(X ,Y); drop(seeds));
report("completed")
```

This solution represents a single thread process (oriented on a single vehicle), where the whole territory is incrementally navigated until all being covered.

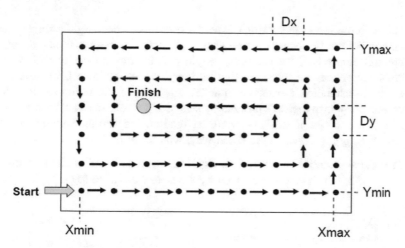

Fig. 11.31 Sequential spiral navigation

- **Sequential spiral, outwards-inwards territory navigation**, as shown in Fig. 11.31, can be accomplished by the following scenario.

```
frontal(
  X = Xmin =..., Xmax = ..., Y = Ymin =..., Ymax =...,
  Dx = ..., Dy = ..., Direction = right);
move(X, Y);
repeat(
  firstrespond(
  (Direction == right;
   if(X + Dx <= Xmax, X += DX,
      (Y += Dy <= Ymax; Ymin += Dy;
       Direction = up))),
  (Direction == left;
   if(X - Dx >= Xmin, X = DX,
      (Y -= Dy >= Ymin; Ymax -= Dy;
       Direction = down))),
  (Direction == up;
   if(Y + Dy <= Ymax, Y += Dy,
      (X -= Dx >= Xmin; Xmax -= Dx;
       Direction = left))),
  (Direction == down;
   if(Y - Dy > Ymin, Y -= Dy,
      (X += Dx <= Xmax; Xmin += Dx;
       Direction = right))));
  move(X,Y); drop(seeds));
report("completed")
```

This solution represents a single thread process too, oriented on a single vehicle as the one before. Choosing the proper solution between these two may depend on different circumstance, for example, terrain peculiarities (where elevation of the region's center may be same as other territory, higher, or lower), weather conditions, final destination advantage, quickest marking of the perimeter as already processed to confirm region's proprietary status, and so on. In a similar way we can organize the opposite, inwards-outwards navigation, like starting from the top of the hill and then going down in the spiral way, which can be less energy consuming.

- **Parallel-sequential territory navigation**, with one of options shown in Fig. 11.32, which can be achieved by the scenario that follows.

```
frontal(
  X = Xmin = ..., Xmax = ..., Y = Ymin = ..., Ymax = ...,
  Dx = ..., Dy = ...);
nodal(DD = 0);
cycle(Y + DD <= Ymax; Y += DD; DD += Dy);
repeat(move(X, Y); drop(seeds);
    X + Dx <= Xmax; X += Dx);
report(completed)
```

For this option, we first created a number of independent starting points at the left of the region using semantics of rule `cycle` (as in the Chap. 4) and then launched individual navigation branch from each of them. These braches operate in parallel, moving through the region from left to right until reach the right edge of the region, each reporting its termination.

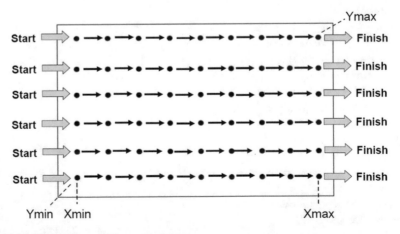

Fig. 11.32 Parallel-sequential area navigation

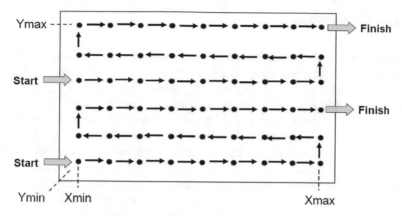

Fig. 11.33 Another combined parallel-sequential navigation

Another variant of parallel-sequential region navigation and processing is shown in Fig. 11.33 with fewer parallel branches, each of which is using horizontal left-to-right then right to left technique described in one of the options above. The SGL solution for this follows.

```
frontal(
  X = Xmin = ..., Xmax = ..., Ymin = ..., Ymax, Y,
  Dx = ..., Dy = ..., Direction = right);
branch(Y = Ymin,
       Y = Ymin + 3 * Dy);
Ymax = Y + 2 * Dy;
repeat(
  or(
    (Direction == right;
     if(X + Dx <= Xmax, X += DX,
       (Y += Dy; Direction = left))),
    (Direction == left;
     if(X - Dx >= Xmin, X -= DX,
       (Y += Dy; Direction = right))));
  Y <= Ymax;
  move(X,Y); drop(seeds)));
report(completed)
```

Any other, more complex, methods of parallel-sequential navigation and processing of the region can be presented in SGL too.

11.12   Conclusion

We have considered different scenarios in SGL for organizing collective unmanned solutions of practical problems with possible use of ground, aerial, surface and underwater robotic units. All solutions appeared to be very compact and can be composed at runtime, with robotic collectives operating fully autonomously, without external control. Taking also into account the C2-oriented scenarios of the previous chapter, which could have both manned and unmanned implementations, we may consider SGT as a step forward in gradual transition to advanced unmanned systems with massive use of cooperative robotics. Providing semantic level descriptions of any spatial tasks within the same formalism allows us to organize any human-robotic teams, where humans and robots can be substituted with each other at runtime, without interrupting the missions.

References

1. LS3—Legged squad support systems. Boston dynamics. http://www.bostondynamics.com/robot_ls3.html
2. CHEETAH—Fastest legged robot. Boston dynamics (2013). http://www.bostondynamics.com/robot_cheetah.html
3. P. Lin, G. Bekey, K. Abney, *Autonomous Military Robotics: Risk, Ethics, and Design* (US Department of Navy, Office of Naval Research). 20 December 2008. http://www.unog.ch/80256EDD006B8954/(httpAssets)/A70E329DE7B5C6BCC1257CC20041E226/$file/Autonomous+Military+Robotics+Risk,+Ethics,+and+Design_lin+bekey+abney.pdf
4. A. McDuffee, DARPA plans to arm drones with missile-blasting lasers. WIRED, 11.01.13. http://www.wired.com/2013/11/drone-lasers/
5. X-47B UCAS, Capabilities, Northrop Grumman, 2015. http://www.northropgrumman.com/Capabilities/X47BUCAS/Pages/default.aspx
6. Meet the SR-72, Lockheed Martin, 2013, http://www.lockheedmartin.com/us/news/features/2013/sr-72.html
7. Engineering Review Board concludes review of HTV-2 second test flight, DARPA, 20 April 2012, http://www.darpa.mil/newsevents/releases/2012/04/20.aspx
8. B. Berkowitz, Sea power in the robotic age. ISSUES in science and technology, 2015, http://issues.org/30-2/bruce-2/
9. Large Displacement Unmanned Underwater Vehicle Innovative Naval Prototype (LDUUV INP), in Naval Drones, http://www.navaldrones.com/LDUUV-INP.html
10. S. Wood, *Autonomous Underwater Gliders* (Florida Institute of Technology). http://my.fit.edu/~swood/26_Wood_first.pdf
11. J. Hsu, U.S. Navy tests robot boat swarm to overwhelm enemies. IEEE spectrum, 5 Oct 2014. http://spectrum.ieee.org/automaton/robotics/military-robots/us-navy-robot-boat-swarm
12. R.L. Hotz, Harvard scientists devise robot swarm that can work together, Wall Street J. 15 Aug 2014. http://www.wsj.com/articles/harvard-scientists-devise-robot-swarm-that-can-work-together-1408039261
13. S. Ljungblad, J. Kotrbova, M. Jacobsson, H. Cramer, K. Niechwiadowicz, Hospital robot at work: something alien or an intelligent colleague? Session: collaboration in the wild, 11–15 Feb 2012, Seattle, WA, USA

14. P. Sapaty, M. Sugisaka, Optimized space search by distributed robotic teams, in *Proceedings of International Symposium on Artificial Life and Robotics (AROB 8th)*, Beppu, Japan, pp. 189–193, 24–26 Jan 2003
15. K. Lobosco, Drones can change the fight against wildfires, 19 Aug 2013, CNN, http://money.cnn.com/2013/08/19/technology/innovation/fire-fighting-drones/
16. P.S. Sapaty, Advanced maritime operations under network-centric organizations, in *Proceedings of the International Conference Warship 2016: Advanced Technologies In Naval Design, Construction and Operation*, 15–16 June 2016, Guildhall, Bath, UK
17. P.S. Sapaty, Advanced naval operations under spatial grasp technology, in *Naval Combat Systems* (Park Plaza Victoria, London), 28–29 July 2015
18. J. Stanley, C. Crump, Protecting privacy from aerial surveillance: recommendations for government use of drone aircraft, American Civil Liberties Union, December 2011. https://www.aclu.org/files/assets/protectingprivacyfromaerialsurveillance.pdf
19. A. Sanchez, C. McKibben, Worst case scenario: the criminal use of drones, Council on hemispheric affairs, Friday, 28 Oct 2016. http://www.coha.org/worst-case-scenario-the-criminal-use-of-drones/
20. J. Paneque-Gálvez, M.K. McCall, B.M. Napoletano, S.A. Wich, L.P. Koh, Small drones for community-based forest monitoring: an assessment of their feasibility and potential in tropical areas, Forests, **51**, 481–1507 (2014)
21. L. Tang, G. Shao, Drone remote sensing for forestry research and practices, J. For. Res. **26** (4):791–797, (2015). https://www.researchgate.net/publication/283655699_Drone_remote_sensing_for_forestry_research_and_practices
22. C. Malveaux, Investigating the potential for drone use in agriculture, Louisiana Agric. **57**(1), 8–11 (2014)

Chapter 12
Conclusions

12.1 General Advantages of the Technology Developed

The Spatial Grasp Technology presented in this book cultivates spatial vision with finding effective solutions on highest possible levels in distributed dynamic systems, allowing at the same time, if needed, to describe, manage, and simulate any other system levels too. SGT can task and re-task complex missions at runtime, on the fly, quickly responding to changing situations, goals, and environments.

Due to formalized presentation of missions in a special high-level language it becomes possible to effectively automate command and control and broadly use unmanned components. Based on full interpretation and unlimited mobility of program code in distributed networked environments, SGT has high level or ubiquity and self-recovery. It can be used in both live and simulation modes as well as any combination thereof.

SGT can effectively establish global control over large distributed systems of any natures. It can use numerous scattered and dissimilar facilities in an integral and holistic way allowing them to work together in a goal-driven supercomputer mode. This parallel supercomputer (more correctly: super-machine) can spatially match arbitrary complex search and processing requests worldwide, producing intelligent resultant knowledge rather than usual huge collection of raw data which may be difficult to comprehend.

12.2 Contribution to the System Theory and Practice

We can show here how different features of the approach developed can contribute to the existing system theories and approaches briefed in Chap. 2, especially in relation to their separate key points (highlighted in italics).

© Springer International Publishing AG 2017

P.S. Sapaty, *Managing Distributed Dynamic Systems with Spatial Grasp Technology*, Studies in Computational Intelligence 690, DOI 10.1007/978-3-319-50461-2_12

General Systems Theory

- *Identical systems principles in various disciplines.* Many system examples of different natures were shown in SGL within the same syntax and semantics, which can confirm that identical system principles really exist and can be expressed by clear and simple constructs and mechanisms (Chaps. 3, 4, 5–7).
- *Wholeness, holism over reductionism.* Practically all examples of creation, analysis, modification and processing of distributed systems have been shown in SGL in a very compact and holistic manner capable of being applied to any systems and with any structures, the latter being the result of these holistic descriptions rather than vice versa (Chaps. 2, 5–8).
- *Open systems.* SGT allows us to create open systems which can evolve at runtime, with new parts added while others removed. The systems will continue preserving high integrity and goal orientation before, during, and after the changes made (Chaps. 7, 10, 11).
- *Application of general system principles to social sciences.* Examples related to social systems with the use of social networks and graph theory have been presented in SGL exhibiting their holistic vision and management, which may be contrary to the existing atomistic conceptions in the field (Chaps. 3, 8, 9).

System Dynamics

- *Complex systems are counterintuitive.* Contribution of SGT here can be with its parallel scenarios spatially grasping the systems as a whole, which can potentially enhance the human intuition. The self-evolving scenarios can reflect different hypotheses of intuitive understanding of large dynamic systems and test them by spatially matching the worlds navigated (Chaps. 7–9).
- *They have a multiplicity of interacting feedback loops.* Such feedback loops can be supported by SGT in different ways. First, by creating arbitrary networked structures, which may be active, it is possible to reflect in them any feedback loops, positive and negative, which will operate as integral parts of the system's body. Second, the spatial SGL patterns, creating and matching these systems, can contain different kinds of higher-level feedback mechanisms themselves, both explicit and implicit, the latter embedded into the automatic language interpretation (Chaps. 4–7).
- *They are controlled by nonlinear relationships.* Any networked structures created in SGL can contain any relations between their elements and parts, including those resulting in nonlinear relationships, which can be effectively studied in a distributed simulation mode by SGT (Chaps. 7–9).
- *Basic elements of network flow models have been introduced.* SGT offers a universal syntax and semantics of basic mechanisms that can create and manage any distributed systems. These include explicit language-based constructs as well as implicit ones invoked automatically during parallel and distributed language interpretation. The latter effectively using the internal, hidden, distributed track system having its own basic elements at nodes and links and operating as "behind the scenes" hierarchical and parallel control-processing

engine. All this fully covers the capabilities of System Dynamics to describe and model large networked systems, while providing much more advanced and universal ones (Chaps. 3–7, 8).

Gestalt Psychology

- *The main concept is unified whole, with elements understood as parts relative to such wholes.* Almost all scenarios for solution of practical tasks in the book represent integral, holistic, descriptions without explicit system parts, the latter emerging automatically during parallel and distributed interpretation of such wholes expressed in SGL and placed into proper environments under certain contexts (Chaps. 2, 3, 7, 8, 10).
- *Mind understands external stimuli as a whole rather than the sum of their parts.* The presented holistic SGL scenarios inherit the capability and experience of human mind to understand and impact the systems as a whole, often regardless of the number of system parts and how they are interconnected. Moreover, operating in highly parallel, distributed and ubiquitous manner, these scenarios may potentially outperform the human mind in its quality to grasp large distributed systems as a whole (Chaps. 7–11).
- *"Gestalt" is usually translated as form, pattern, although it might be better understood as organized structure.* All presented SGL scenarios are based on different patterns self-matching distributed systems, which can be enriched with arbitrary compositions of universal constructs called rules, altogether providing organized structures of high cognitive and operational powers (Chaps. 3, 4, 10, 11).
- *The wholes are structured and organized using grouping laws, called laws or principles.* These can be easily expressed as certain patterns in SGL and used to find related images in both virtual and physical environments. The existing laws, however, had been formulated mostly for visual images and are by no means sufficient for understanding and management of large distributed systems. They need considerable extension and diversification to be of practical use in this area, and the exercised pattern matching examples in this book can hint on possible ways towards more general and universal grouping laws oriented on distributed systems.

Memetics

- *The meme, analogous to a gene, as a unit of culture is hosted in the minds of one or more individuals.* There are virtually unlimited possibilities of representing memes within SGL syntax—from elementary cultural units to their arbitrary structured complexes, or "memeplexes", which can self-replicate and move in space in parts or as a whole (Chaps. 4, 6–8).
- *They spread through the social body similar to how genes spread through the biological body.* This can be effectively implemented and modelled in SGL, as memes and their complexes can move with waves-based self-propagating patterns in frontal variables when navigating networked systems, which may represent social networks too (Chaps. 3, 6, 7, 9).

- *Memes can reproduce themselves, like a virus moves from body to body, they move from mind to mind.* Unlimited broadcasting and self-replicating capability of SGL scenarios can perfectly reflect and model these features, where self-evolving and self-propagating patterns can carry replicated memes and memeplexes on parallel wave fronts in automatically replicated frontal variables, in a virus-like manner. Any transformation, modification, grouping or splitting of the propagating memes under any conditions for doing so can be expressed in SGL patterns self-spreading in distributed worlds (Chaps. 7–9).
- *Memetics can be effectively used to identify and target root causes of social problems.* SGT allows us to undertake thorough investigation of any such problems in real and very large social networks (extending throughout any territories up to the whole world) with the use of all associated with these networks computational and information resources working cooperatively under SGT and activated by the spreading parallel scenarios in SGL (Chaps. 6–9).

Brain Waves and Consciousness

- *Brainwaves reflect our state of mind.* SGT, in its infancy, was strongly inspired by brainwaves, with original language WAVE based on coordinated wavelike navigation of distributed systems rather than representing solutions in them as cooperating parts or agents. This allowed us to get integral and compact solutions of many graph and network problems, where each graph node could potentially reside in a different computer, also spread waves of active code via the internet and between different countries. The latest language version, SGL, uses similar wave ideology but only as one of the technology features, the others being powerful feedback and echo processing mechanisms triggered by SGL rules and operating automatically and implicitly on the level of internal language implementation, whereas in WAVE they had to be programmed explicitly while making programs more complex and less transparent. Waves in SGT (in WAVE too) are active substances bringing activity to distributed systems which may be passive, being exactly upside down as compared to the nature of brain waves, which are entirely the result of the system's (i.e. brain's) activity (Chaps. 1, 3, 8).
- *Each brainwave has its own set of characteristics representing a unique state of consciousness.* Similarly to this, the activity and frequency of the self-spreading waves by SGL scenarios can influence the spatial awareness, productivity, integrity and, if you do not mind, "consciousness" of the distributed solutions and their ability to update themselves and timely respond to rapidly changing situations in dynamic systems (Chaps. 1, 3, 7, 11).
- *The concept of waves is also used in a study of higher levels of mental activity like, for example, in integral psychology.* The wavelike representation and evolution of SGL scenarios may also have psychological analogy, in how we think of and percept distributed spaces while shifting attention and concentration from one points and details to the other, which may be in combination with

direct parallel perception of the whole visual image by the brain, which is also reflected in SGL (Chaps. 3, 4, 8, 10).

- *Mind and brain need to be included in a non-reductionistic way in any integral theory of consciousness.* Many SGL scenarios presented in the book do not separate explicitly the system creation from the system supervision, evolving altogether as an integral spatial process. This highlights holistic, non-reductionistic SGT nature of dealing with distributed systems, treating their "mind" and "brain" within the same spatial grasp concept (Chaps. 3, 4, 6–8, 11).

Interoperability and Over-operability

- *Different kinds of interoperability: syntactic interoperability, semantic interoperability, cross-domain interoperability.* SGT can effectively support these and any other, far more advanced interoperability principles. It can, for example, provide distributed cooperation where interaction takes place not only through dedicated points and interfaces but rather in the form of "overlapping fields". Each such field representing a separate, possibly, dynamically created organization, with their any depth and breadth intersection (penetrating into each other's information and control domains)—up to a full merge into the united, single, system (Chaps. 3, 7, 10, 11).
- *Over-operability versus interoperability for High-Level System Management.* This can be organized by higher-level patterns both creating and covering distributed systems, with potentially unlimited tasking and supervision capabilities expressed in SGL. The same language expression of controlling and controlled systems allows different spatial processes penetrate into each other's information and control fields and infrastructures and find holistic goal-driven solutions (Chaps. 3, 7, 10, 11).

12.3 Some Particular Application Areas

SGT can analyze and properly impact critical infrastructures, both native and adversary's, as well as create new infrastructures for a variety of purposes. Under SGT, multiple targets can be simultaneously captured over large areas and individually traced & analyzed by spreading mobile intelligence. Based on full interpretation of SGL scenarios capable of re-launching their parts or the whole at runtime, the distributed systems empowered with SGT can remain fully operational after indiscriminate damages of arbitrary components. These and many other capabilities of the approach described in the book may have many useful applications in most diverse system fields. Of particular importance may be the following application areas for which SGT was successfully reported at different world events (some already mentioned and discussed in different chapters while others can be obtained by the provided publication references).

- *Distributed interactive simulation* [1–6]. The SGT (as former WAVE too) had been researched for both live control of large dynamic systems like battlefields and distributed interactive simulation of them (the latter serving as a look-ahead to the former), also any combination thereof, with watershed between the two changing at runtime. The approach was used in the past as part of the international Distributed Interactive Simulation (DIS) project headquartered in Orlando, Florida.
- *Intelligence, Surveillance and Reconnaissance (ISR)* [7–10]. SGT can integrate distributed ISR facilities into flexible goal-driven systems operating under unified command and control, which can be automatic. These integrated systems, being empowered with spatial intelligence in SGL, can survey and analyze distributed situations extracting high-level knowledge from huge amounts of observable raw data, also properly impact critical infrastructures, both native and adversary's, as well as create new infrastructures for a variety of purposes.
- *Tactical Communications* [11–13]. Tactical communications are basic in military operations, and SGT can effectively put them on a much higher, semantic, level effectively integrating with advanced command and control. This can drastically reduce amounts of data to be exchanged between system components especially in time and security critical situations, also when communications cannot proceed with normal speed like, for example, in underwater operations.
- *Electronic Warfare (EW)* [14–16]. SGT can behave as a powerful intelligent virus self-spreading through networked systems, the latter having numerous SGL interpreters embedded into their infrastructures (in stealth manner, if required). This super-virus can set up any control over the navigated systems and incapacitate any other EW facilities, also effectively block, fight, and eliminate malicious computer viruses by tracing their sources in networks.
- *Directed Energy (DE) Systems* [17, 18]. The networked DE systems can be programmed in SGL on the fly timely reacting on the situations occurred and securing delivery of directed energy to any points of the globe though air and space (using multiple satellite mirrors for the space-based operations). SGT can also provide flexible tactical solutions integrating of DE systems with advanced robotics.
- *Global Awareness and Dominance* [19–25]. SGT can establish overwhelming dominance over distributed networked systems. Self-evolving and self-spreading SGL scenarios can collect and generalize distributed data, spreading and sharing their findings throughout the controlled systems, thus providing global awareness of distributed situations at any point. They can dynamically create different kinds of operational infrastructures (both open and stealthy) or jam and eliminate such in adversary's environments.
- *Massive Robotics* [26–34]. SGT paves the way for unified transition to automated up to fully unmanned systems with massive use of advanced robotics. One of practical benefits may be the effective management of advanced robotic collectives, like drone swarms, regardless of their size and spatial distribution and by only a single human operator.

- *System Integrity* [27, 35–37]. SGL can integrate any system resources and any other systems into holistic systems of systems that can pursue common goals and react in an optimal way on emerging threats, globally optimizing the use of scattered and possibly scarce resources, say, to withstand massive missile attacks in an optimum way, if this ever happens.
- *Sensor Networks* [38–40]. Multiple sensors scattered over large territories can behave under SGT as a spatial supercomputer operating under local and even unsafe and restricted communications, also without any central control, but effectively pursuing global goals, where individual sensor units can be stationary or mobile.
- *Command and Control* [41, 42]. Description in SGL of semantic-level military missions is much clearer and more compact (up to 10 times) than if written in existing battle management languages. This simplicity may allow us to redefine the whole scenario or its parts at runtime when goals and environment change rapidly, especially for asymmetric situations and operations, also engage robotic units most naturally and efficiently.
- *Human terrain* [43]. SGT allows this new field, originally coined in military, to be considered and used in a much broader sense and scale than initially planned, allowing it to solve complex national and international conflicts and problems by intelligent, peaceful and predominantly nonmilitary means while obeying existing ethical standards.
- *Air and Missile defense* [44, 45]. Providing flexible and self-recovering distributed C2 infrastructures, SGT can organize distributed defense systems withstanding multiple ballistic missiles and recovering at runtime after damages. It also can, for example, effectively use distributed networks of cheap ground or low-altitude sensors to discover, trace, and destroy multiple cruise missiles with complex routes, versus existing expensive high-altitude planes, drones, and aerostats.
- *Crisis management* [46, 47]. SGT can effectively support advanced distributed systems for crisis management operations, where complex relief missions, national and international, can be organized and programmed in SGL quickly, "on the fly", with the use of any available resources, human as well as robotic.
- *Night Vision* [48]. Under SGT, multiple communicating sensors distributed throughout any regions can operate together as an integral spatial brain providing holistic vision of the operational theatre. This can help to continue solving many important tasks, surveillance and reconnaissance including, even at night or in bad weather, compensating insufficiency of sensor data with enhanced intelligence of the whole system.

Many application ideas and examples mentioned above were presented and discussed at a number of invited presentations at different military-oriented international conferences with the main goal to *test the developed Spatial Grasp Technology in most severe conditions and situations*. The experience gained from these meetings helped enormously to cultivate holistic vision of dynamic distributed systems and improve Spatial Grasp Language as the core of SGT, having enriched the language with powerful practical constructs which would be unimaginable from purely theoretical analysis or existing civil technologies and applications. SGT, however, is planned *to be widely used in predominantly civil areas bringing to them new methods tested with defense.* Its extended marketing, especially in advanced infrastructure operations and massive robotics, is expected too.

12.4   Implementation Issues and Future Plans

The described approach had been prototyped and tested, via internet including, in different countries and on a variety of networked applications. Whereas interpreters of the previous SGL versions, like WAVE, were demonstrated with special communication channels and protocols, the current SGL interpreter can use for communication with other interpreters any existing channels like, for example, electronic mail (updated to operate in automatic mode) or any other media ones.

This can be organized even within existing university environments and by a small group of students who may get interested in networking and distributed processing and control, especially in relation to their M.Sc. and Ph.D. projects, as was done in the past at different universities in Germany, UK, Canada, and Ireland. The author will be happy to provide any needed assistance in this respect.

A new patent on the described technology is being prepared, another book too, the latter on more fundamental issues of distributed systems theory and practice with their holistic, gestalt-based, supervision and management.

References

1. P.S. Sapaty, M.J. Corbin, P.M. Borst, Mobile WAVE programming as a basis for distributed simulation and control of dynamic open systems. Report at the 4th UK SIWG national meeting, SGI Reality Centre, Theale, Reading, 11 Oct 1994
2. P.S. Sapaty, M.J. Corbin, P.M. Borst, *Towards the Development of Large-Scale Distributed Simulations.* Proceedings of 12th Workshop on Standards for the Interoperability of Distributed Simulations, IST UCF, Orlando, FL, Mar 1995, pp. 199–212
3. P.S. Sapaty, M.J. Corbin, S. Seidensticker, *Mobile Intelligence in Distributed Simulations.* Proceedings of 14th Workshop on Standards for the Interoperability of Distributed Simulations, IST UCF, Orlando, FL, Mar 1995
4. P.S. Sapaty, P.M. Borst, M.J. Corbin, J. Darling, *Towards the Intelligent Infrastructures for Distributed Federations.* Proceedings of 13th Workshop on Standards for the Interoperability of Distributed Simulations, IST UCF, Orlando, FL, Sept 1995, pp. 351–366
5. P.S. Sapaty, *A New Technology for Integration, Simulation, and Testing of Distributed Dynamic Systems.* NATO Proceedings "Integration of Simulation with System Testing", RTO-MP-083, AC/323(SCI-083)TP/43, June 2002, 12 p
6. P.S. Sapaty, Distributed interactive simulation and control of collective aerial operations. International conference military flight training, London, UK, 18–19 Sept 2013. http://www.smi-online.co.uk/defence/uk/military-flight-training

7. P.S. Sapaty, Integration of ISR with advanced command and control for critical mission applications. SMi's ISR conference, Holiday Inn Regents Park, London, 7–8 Apr 2014
8. P.S. Sapaty, Providing over-operability of advanced ISR systems by a high-level networking technology. SMI's airborne ISR, Holiday Inn Kensington Forum, London, United Kingdom, 26–27 Oct 2015
9. P.S. Sapaty, Organization of advanced ISR systems by high-level networking technology. MMC, No 1 (2016)
10. P.S. Sapaty, Integral spatial intelligence in ISR applications. SMi's ISR workshop, Holiday Inn Regents Park London, 9 Apr 2014
11. P.S. Sapaty, *High-Level Communication Protocol for Dynamically Networked Battlefields.* Proceedings of International Conference Tactical Communications 2009 (Situational Awareness and Operational Effectiveness in the Last Tactical Mile), One Whitehall Place, Whitehall Suite and Reception, London, UK (2009)
12. P.S. Sapaty, *Tactical Communications in Advanced Systems for Asymmetric Operations.* Proceedings of Tactical Communications 2010, CCT Venues, Canary Wharf, London UK, 28–30 Apr 2010
13. P.S. Sapaty, *Emerging Asymmetric Threats, Q&A Session.* Tactical Communications 2010, CCT Venues, Canary Wharf, London UK, 28–30 Apr 2010
14. P.S. Sapaty, Distributed capability for battlespace dominance. Electronic warfare 2009 conference and exhibition, Novotel London West Hotel and Conference Center, London (2009)
15. P.S. Sapaty, Global electronic dominance with spatial grasp. Int. J. Commun. Networks Syst. Sci. **5**(11) (2012, November)
16. P.S. Sapaty, *Global Management of Distributed EW-Related System.* Proceedings of International Conference Electronic Warfare: Operations & Systems 2007, Thistle Selfridge, London, UK, 19–20 Sept 2007
17. P.S. Sapaty, A. Morozov, M. Sugisaka, *DEW in a Network Enabled Environment.* Proceedings of the International Conference Directed Energy Weapons 2007, Le Meridien Piccadilly, London, UK, 28 Feb–1 Mar 2007
18. P.S. Sapaty, *High-level Organisation and Management of Directed Energy Systems.* Proceedings of Directed Energy Weapons 2010, CCT, Canary Wharf, London UK, 25–26 Mar 2010
19. P.S. Sapaty, *Distributed Technology for Global Dominance.* Proceedings of International Conference Defense Transformation and Net-Centric Systems 2008, as part of the SPIE Defense and Security Symposium, World Center Marriott Resort and Convention Center, Orlando, FL, USA, 16–20 Mar 2008 (Proceedings of SPIE—Volume 6981, Defense Transformation and Net-Centric Systems 2008, Raja Suresh, Editor, 69810T, Apr. 3, 2008)
20. P.S. Sapaty, *Distributed Technology for Global Dominance.* Keynote lecture, Proceedings of the Fifth International Conference in Control, Automation and Robotics ICINCO 2008, The Conference Proceedings, Funchal, Madeira, Portugal, 11–15 May 2008
21. P.S. Sapaty, Distributed capability for battlespace dominance. Electronic warfare 2009 conference and exhibition, Novotel London West Hotel and Conference Center, London, 14–15 May 2009
22. P.S. Sapaty, *Distributed Technology for Global Control.* Book chapter, Lecture Notes in Electrical Engineering, Informatics in Control, Automation and Robotics, vol. 37 (Springer, Berlin, 2009)
23. P.S. Sapaty, Global electronic dominance. 12th international fighter symposium, Grand Connaught Rooms, London, UK, 6–8 Nov 2012
24. P.S. Sapaty, Providing global awareness in distributed dynamic environments. International summit ISR, London, 16–18 Apr 2013. http://www.smi-online.co.uk/defence/uk/isr
25. P.S. Sapaty, Ruling distributed dynamic worlds with spatial grasp technology. Tutorial at the international science and information conference 2013 (SAI), London, UK, 7–9 Oct 2013
26. P.S. Sapaty, Towards massively robotized systems under spatial grasp technology. J. Comput. Sci. Syst. Biol. **9**(1) (2016)

27. P.S. Sapaty, *Providing Spatial Integrity for Distributed Unmanned Systems*. Proceedings of 6th International Conference in Control, Automation and Robotics ICINCO 2009, Milan, Italy (2009)
28. P.S. Sapaty, Unified transition to cooperative unmanned systems under spatial grasp paradigm. Int. J. Trans. Networks Commun. **2**(2) (2014, April)
29. P.S. Sapaty, Military robotics: latest trends and spatial grasp solutions. Int. J. Adv. Res. Artif. Intell. **4**(4) (2015)
30. P.S. Sapaty, From manned to smart unmanned systems: a unified transition. SMi's military robotics, Holiday Inn Regents Park London, 21–22 May 2014
31. P.S. Sapaty, *High-Level Technology to Manage Distributed Robotized Systems*. Proceedings of Military Robotics 2010, Jolly St Ermins, London, UK, 25–27 May 2010
32. P.S. Sapaty, Human-Robotic Teaming: A Compromised Solution. AUVSI's Unmanned Systems North America 2008, San Diego, USA, 10–12 June 2008
33. P.S. Sapaty, M. Sugisaka, *Distributed Artificial Brain for Collectively Behaving Mobile Robots*. Proceedings of Symposium and Exhibition Unmanned Systems 2001, Baltimore, MD, Jul 31–Aug 2 2001, 18 p
34. P.S. Sapaty, Unified Transition to Robotized Armies with Spatial Grasp Technology. International Summit Military Robotics, London, United Kingdom, 12–13 Nov 2012
35. P.S. Sapaty, Towards wholeness and integrity of distributed dynamic systems. J. Comput. Sci. Syst. Biol. **9**(3) (2016)
36. P.S. Sapaty, Towards global goal orientation, robustness and integrity of distributed dynamic systems. J. Int. Relat. Diplomacy **4**(6) (2016, June)
37. P.S. Sapaty, *Gestalt-Based Integrity of Distributed Networked Systems*. SPIE Europe Security + Defence, bcc Berliner Congress Centre, Berlin, Germany (2009)
38. P.S. Sapaty, *Remote Control of Open Groups of Remote Sensors*. Proceedings of SPIE Europe Security + Defence, Berlin, Germany (2009)
39. P.S. Sapaty, M. Sugisaka, M.J. Delgado-Frias, J. Filipe, N. Mirenkov, Intelligent management of distributed dynamic sensor networks. Artif. Life Robot. **12**(1–2), 51–59 (2008, March)
40. P.S. Sapaty, M. Sugisaka, J. Filipe, *Making Sensor Networks Intelligent*. Proceedings of the 4th International Conference on Informatics in Control, Automation and Robotics, ICINCO-2007, Angers, France, 9–12 May 2007
41. P.S. Sapaty, Unified transition to cooperative unmanned systems under spatial grasp paradigm. 19th international command and control research and technology symposium, Alexandria, Virginia, 16–19 June 2014
42. P.S. Sapaty, Formalizing commander's intent by spatial grasp technology. Accepted paper at the international society of military sciences (ISMS) 2012 annual conference, Kingston, Ontario, Canada, 23–24 Oct 2012
43. P.S. Sapaty, Distributed human terrain operations for solving national and international problems. Int. Relat. Diplomacy **2**(9) (2014, September)
44. P.S. Sapaty, Distributed missile defence with spatial grasp technology. SMi's military space, Holiday Inn Regents Park London, 4–5 Mar 2015
45. P.S. Sapaty, Distributed air and missile defense with spatial grasp technology. Intell. Control Autom. Sci. Res. **3**(2), 117–131 (2012)
46. P.S. Sapaty, Crisis management with distributed processing technology. Int. Trans. Syst. Sci. Appl. **1**(1), 81–92. ISSN 1751-1461 (2006)
47. P.S. Sapaty, M. Sugisaka, R. Finkelstein, J. Delgado-Frias, N. Mirenkov, *Emergent Societies: An Advanced IT Support of Crisis Relief Missions*. Proceedings of Eleventh International Symposium on Artificial Life and Robotics (AROB 11th'06), Beppu, Japan, 23–26 Jan 2006, ISBN 4-9902880-0-9
48. P.S. Sapaty, Night vision under advanced spatial intelligence: a key to battlefield dominance. International summit night vision 2013, London, 4–6 June 2013